Critical Thinking and Creative Analogies in Statistics, Science, and Technology

Through the lens of critical thinking and creative analogy, this book skillfully blends mainstream perspectives with bold, thought-provoking personal insights, offering readers a fresh and engaging perspective on complex topics. By leveraging critical thinking, creative analogies, and practical examples from statistics, medicine, socioeconomics, education, and technology, it bridges the gap between abstract theory and real-world applications.

Each chapter is concise and impactful, cutting straight to the essence of the subject. Thought experiments and vivid examples illuminate key concepts, making them both accessible and actionable. Whether you're seeking clarity, inspiration, or a deeper understanding, this book delivers powerful, thought-provoking content that will leave a lasting impression.

Key Features

- A harmonious balance of mainstream views and provocative personal insights.
- Creative analogies paired with practical examples from medicine and other fields.
- Concise, clear, and practical chapters that focus on core ideas, enriched with thought experiments and real-world applications.
- A progressive approach, moving from simple daily decision-making to the development of integrated, humanized AI.
- Chapter exercises designed to reinforce concepts through hands-on practice.

Mark Chang, Ph.D., is the founder of AGInception and an Adjunct Professor of Biostatistics at Boston University. He is an elected fellow of the American Statistical Association, a co-founder of the International Society for Biopharmaceutical Statistics. He has previously held various positions in pharmaceutical companies, including Scientific Fellow, Executive Director, and Senior Vice President. He is an adaptive design expert and has extensive knowledge in AI for clinical trials and humanized AI. He has published 14 books on artificial intelligence and machine learning, adaptive clinical trial designs, modern issues and methods in biostatistics, paradoxes in scientific inference, and principles of scientific methods.

"The AI age is defined by the value of questions over answers, adaptability over stability, and connections over isolated knowledge."

Critical Thinking and Creative Analogies in Statistics, Science, and Technology

Essential Skills for the AI Era

Mark Chang

CRC Press
Taylor & Francis Group
Boca Raton London New York

CRC Press is an imprint of the
Taylor & Francis Group, an **informa** business

A CHAPMAN & HALL BOOK

Designed Cover: Shutterstock

First edition published 2026
by CRC Press
2385 NW Executive Center Drive, Suite 320, Boca Raton FL 33431

and by CRC Press
4 Park Square, Milton Park, Abingdon, Oxon, OX14 4RN

CRC Press is an imprint of Taylor & Francis Group, LLC

© 2026 Mark Chang

ISBN: 9781041048138 (hbk)
ISBN: 9781041043829 (pbk)
ISBN: 9781003630081 (ebk)

DOI: 10.1201/9781003630081

Typeset in Palatino
by codeMantra

Contents

Preface

With the rapid advancements in AI and machine learning, we stand at a pivotal moment in how knowledge is understood and applied across disciplines. This transformation calls for a rethinking of education and problem-solving approaches in statistics, science, and technology. This book, *Critical Thinking* and *Creative Analogies* *in Statistics, Science, and Technology—Essential Skills in the AI Era,* responds to this need by advocating for a more integrated, creative approach that leverages the power of critical thinking and analogies to navigate an increasingly complex landscape.

The future demands professionals who can seamlessly connect traditional skills with emerging technologies, moving fluidly across domains. Education must go beyond isolated subjects, emphasizing critical thinking, interdisciplinary problem-solving, and the ability to draw connections between seemingly unrelated areas. In statistics, the boundaries between classical methods and machine learning are fading, requiring statisticians to master not only frequentist and Bayesian approaches as well as AI techniques. More crucially, they must harness insights from diverse fields by drawing analogies, enabling innovative solutions to complex, data-driven challenges.

This book explores the role of analogies in addressing a wide range of challenges: resolving the intransitivity of statistical nonparametric tests (A is better than B, B better than C, and C better than A, probabilistically), transforming ineffective drugs into effective treatments through optimal randomization inspired by the fair game paradox, and enhancing clinical trial designs using cooperative game theory. It examines designing trials with drug combinations through Braess's paradox, simplifying statistical modeling with dimensional analysis, optimizing adaptive trials via Brownian motion, and identifying effect sizes without unblinding clinical trial data while addressing various biases using critical thinking. This book further explores applying network analysis to study the connotations of understanding, utilizing play-the-winner analogies to reduce polarization in social systems, employing hierarchical analogies to demonstrate equal probabilities of evolution and devolution, and applying the similarity principle from machine learning to clinical trials and beyond. Later chapters delve into the use of complex analogies in AI and the development of humanized AI. While rooted in biostatistical challenges, this book demonstrates the interdisciplinary power of analogical reasoning through examples spanning diverse fields.

Analogies have shaped the course of history, from early scientific breakthroughs to modern technological innovations. They are the unseen threads connecting disparate ideas, enabling us to weave a tapestry of knowledge that transcends boundaries. More than just cognitive tools, analogies reflect our capacity for imagination and insight, offering a powerful framework for understanding, problem-solving, and creative expression.

Creative analogies exhibit distinctive features, such as identifying problems through outcome reflection, applying solution models from one domain to another, and linking specific challenges to universal principles. They also reveal structural parallels, operate hierarchically or recursively, and connect ideas through the principle of similarity—similar entities yielding comparable outcomes as their resemblance increases. These characteristics make analogies invaluable in fostering innovation and addressing the complexities of modern life.

This book intentionally presents perspectives that are provocative, counterintuitive, or challenge conventional wisdom. It often provides fresh perspectives on issues and problems, though this does not necessarily reflect my complete agreement with those views. It begins with simple analogies in everyday decisions and medical contexts, gradually progressing to advanced applications such as bionics in AI and the creation of more human-like intelligence. By starting with paradoxes and dilemmas and building toward complex analogies, this book offers a structured exploration of analogy's transformative power.

Because this book delves into critical thinking and creative analogies, it covers an unusually broad range of fields, encouraging readers to step beyond their comfort zones. At the same time, the chapters are relatively independent, and some sections include advanced mathematics for those who enjoy it. However, if math isn't your preference, you can easily skip those parts without detracting from the overall experience.

While the examples in this book reflect my own perspectives, they are intended as a foundation to inspire further exploration and spark new ideas. Each reader brings unique experiences to the use of creative analogies, enriching the possibilities for discovery and insight.

I have included exercises at the end of each chapter to support professors who may wish to use this book as a primary textbook for a new course in the AI age or as supplementary material for an existing course. The exercises include the following five aspects: understanding key concepts, critical thinking and application, analogy and creativity, debate and discussion, and exploring extensions. I would be delighted to receive their feedback on both the book and the exercises. At the same time, I encourage students to approach these exercises with an open mind. Do not let your imagination be confined by the information presented in this chapter, as traditional exercises often do. Instead, view these prompts as opportunities to think creatively and explore ideas beyond the boundaries of the text through the power of analogy.

Mark Chang
February 2025
Boston University, MA, USA

1

Inventive Connections:
A Journey through Analogies

The AI age is defined by the value of

questions over answers,

adaptability over stability, and

connections over isolated knowledge.

1.1 The Importance of Critical Thinking and Creative Analogy

Critical thinking and creative analogy are vital intellectual tools for problem-solving, innovation, and understanding complex concepts. While they serve different purposes, their interplay enhances our ability to navigate challenges and generate novel ideas, especially in the age of artificial intelligence (AI).

1.1.1 Critical Thinking: The Foundation of Reasoning

Critical thinking is the process of objectively analyzing and evaluating information to form a judgment. It involves questioning assumptions, identifying biases, and drawing logical conclusions.

Critical thinking can be helpful in different ways:

1. Decision-making: Critical thinking enables individuals to assess options systematically and make informed choices, whether in personal decisions, business strategy, or policy-making.
2. Problem-solving: By dissecting complex problems into manageable components, critical thinking allows for methodical solutions.
3. Avoiding cognitive biases: It helps counteract biases like confirmation bias or availability heuristic, ensuring conclusions are based on evidence rather than preconceived notions.

DOI: 10.1201/9781003630081-1

4. Interdisciplinary application: Critical thinking fosters the ability to synthesize knowledge across fields, promoting holistic understanding and cross-disciplinary insights.

1.1.2 Creative Analogy: The Bridge to Innovation

Creative analogy involves drawing comparisons between seemingly unrelated domains to generate new insights or solve problems. It leverages familiar concepts to shed light on unfamiliar or complex ideas.

Analogy offers various benefits, including:

1. Simplifying complexity: Analogies translate abstract or technical concepts into relatable terms, making them easier to understand (e.g., likening the brain to a computer).
2. Fostering innovation: Analogies connect disparate ideas, sparking creative solutions (e.g., Velcro inspired by burrs sticking to fabric).
3. Encouraging empathy: They help individuals see problems from new perspectives, fostering deeper understanding and collaboration.
4. Enhancing communication: Analogies are powerful storytelling tools, helping convey complex ideas to diverse audiences.

1.1.3 Interplay between Critical Thinking and Creative Analogy

While critical thinking ensures logical rigor, creative analogy fuels imaginative leaps. Their combination is transformative and can be illustrated as follows.

1. Enhancing problem-solving: Critical thinking ensures a structured approach, while analogy introduces novel perspectives. For example, using the flow of water in rivers as an analogy to optimize traffic systems, with critical thinking validating the feasibility of the solution.
2. Cross-disciplinary insights: Analogies encourage exploration across fields, while critical thinking assesses the validity of these connections. For example, drawing parallels between ecosystems and organizational behavior to design sustainable business practices.
3. Improving teaching and learning: Analogies help explain concepts, and critical thinking ensures they are applied appropriately. For example, teaching electrical circuits by comparing them to water pipes while critically evaluating where the analogy breaks down.

However, there are challenges and pitfalls when using critical thinking and creative analogy, including:

1. Overreliance on Analogies: Analogies can oversimplify or mislead if critical thinking doesn't identify their limitations.

2. Cognitive rigidity: Excessive focus on logic can stifle creativity if analogical thinking is not encouraged.
3. Balancing the two: Effective problem-solving requires a balance of structured reasoning and imaginative exploration.

In summary, critical thinking and creative analogy are complementary skills that, when used together, enhance our intellectual toolkit. Critical thinking provides the structure and rigor necessary for reliable conclusions while creative analogy fosters innovation and a deeper understanding of complex ideas. Together, they enable us to tackle problems, communicate effectively, and push the boundaries of knowledge and creativity. In a rapidly evolving world, cultivating these skills is essential for personal and professional success.

1.1.4 Similarity: Linking Imitation, Analogy, and Creativity

Imitation is the foundation of learning, creativity, and innovation. All knowledge and wisdom are built on what others have created; we never create something from nothing. Imitation helps us master core ideas and basic skills, forming the bedrock of creativity. Socially, imitation shapes norms and preserves the collective wisdom of generations as culture.

Creativity often begins with inspiration from the existing ideas, implying partial imitation or analogy. An analogy compares two systems or objects, highlighting similarities, and analogical reasoning uses these connections to infer new relationships or solve problems. Analogies are supported by the Similarity Principle, which allows reasoning from shared properties to new insights.

As a form of logic, analogy differs from deduction, induction, and abduction, focusing on particular similarities rather than general rules. It plays a crucial role in cognition, aiding problem-solving, decision-making, memory, perception, creativity, and communication. Analogy is pervasive in ordinary language, science, philosophy, and the humanities, making it essential for developing intelligent agents.

Holyoak and Thagard (1995) define three key characteristics of a good analogy:

1. Similarity: Source and target share common properties.
2. Structure: A systematic correspondence exists between elements of the source and target.
3. Purpose: Analogies are guided by goals and adapt as new information arises.

Creativity connects to imagination and originality, while innovation translates creative ideas into action. Innovations often arise from clever analogies, borrowing across disciplines to produce inventive solutions.

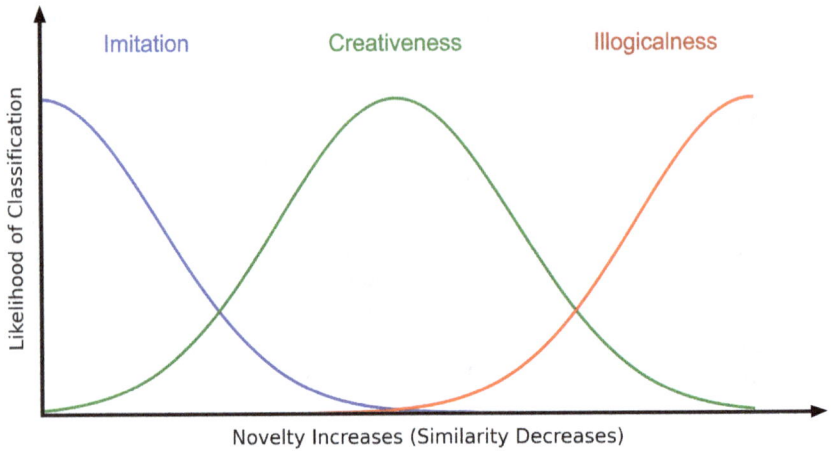

FIGURE 1.1
Imitation, creativity, and illogicality across similarity levels.

Imitation, analogy, and innovation share degrees of similarity: Extreme likeness defines imitation, partial likeness defines analogy, and principled likeness defines innovation. Beyond these scales, ideas with little similarity may seem "crazy" or "illogical." Figure 1.1 illustrates this continuum: imitation operates on great similarity, analogy bridges imitation and creativity, and excessive novelty risks illogicality.

Learning occurs through imitation and analogy. Imitation emphasizes similarity, while analogy evolves imitation by abstracting commonalities between dissimilar things. Creativity emerges when analogies push boundaries, connecting ideas with minimal similarity. Illogicality, by contrast, results when analogies lack sufficient commonality. In our HAI architecture (Chapter 19), imitation, analogy, and creativity coexist under the umbrella of similarity, balancing exploitation and exploration for effective learning and innovation.

1.2 The Paradoxes of Happiness: Critical Reflections on Choices, Technology, and Progress

Our efforts in life often center on achieving happiness and longevity, goals that seem universally appealing. Yet, the relationship between effort and happiness is not straightforward. Hard work does not always lead to the happiness we desire, and some argue that happiness arises not from striving but from the absence of striving. Happiness is deeply intertwined with health, prosperity, and societal factors such as technological advancements, changes in wealth, knowledge, and education. It is also relative: one's happiness often depends on comparisons to others' wealth or social status.

Happiness is influenced by the gap between what one wants and what one gets. The greater the expectations, the harder they are to satisfy, potentially leading to dissatisfaction. As the saying goes, "If you want what you get, you will get what you want." Happiness also stems from having hopes, dreams, and a sense of empowerment. An individual's baseline happiness is shaped by their outlook on life—whether optimistic or pessimistic. Some view the purpose of life as maximizing happiness; others prioritize longevity or striking a balance between the quality and quantity of life. For some, happiness is about enjoying the process of living rather than the outcomes.

Despite its complexity, many believe technological innovation inherently brings happiness. But does it? We often develop technology to save time and make life easier, yet paradoxically, these innovations often lead us to work harder and longer. The time saved by technological advances is frequently reinvested in tackling more challenging tasks or developing even more advanced technologies. Instead of enjoying the leisure these innovations promise, we risk becoming slaves to them. The key question is not how much time we save but how we choose to use that saved time.

The power of choice is another cornerstone of happiness. Having choices makes us feel empowered, yet an abundance of options can lead to "informational obesity" and analysis paralysis. Even after making a good choice, we may feel a sense of loss for the unchosen alternatives. To mitigate this, we have developed AI tools to make recommendations, such as when shopping online. However, AI raises further questions: does it bring convenience and help us make better choices, or does it strip us of certain freedoms, leaving us less satisfied? The answer to this, as with many things, depends on the individual.

This complexity is well captured by Fredkin's Paradox (or Minsky's Optimization Paradox): The more equally attractive two options appear, the harder it becomes to choose between them—even though, paradoxically, the choice matters less. As Marvin Minsky (1986) observed, decision-making agents might spend disproportionate amounts of time on less significant decisions. Instrumental rationality—the pursuit of all means necessary to achieve a specific goal—often overlooks the cost of excessive deliberation. An intuitive solution to Fredkin's Paradox is to treat decision-making time as a cost itself (Klein, 2001), highlighting a tension between rationality and practicality.

While technological innovation has driven exponential societal prosperity, it has also contributed to widening wealth gaps. This dynamic reflects a competitive, micro-level motivation where individuals act in their self-interest, often leading to macro-level outcomes that no one desires. Braess's Paradox illustrates this well: adding more roads to a network can worsen traffic. Similarly, technological advancements can exacerbate inequality, creating unintended societal consequences despite individual gains.

In our pursuit of happiness, wealth, and progress, we must consider these paradoxes and reflect on the broader implications of our choices and innovations. Happiness is not just about achieving goals or accumulating wealth but about balancing long-term aspirations with short-term joys, and individual

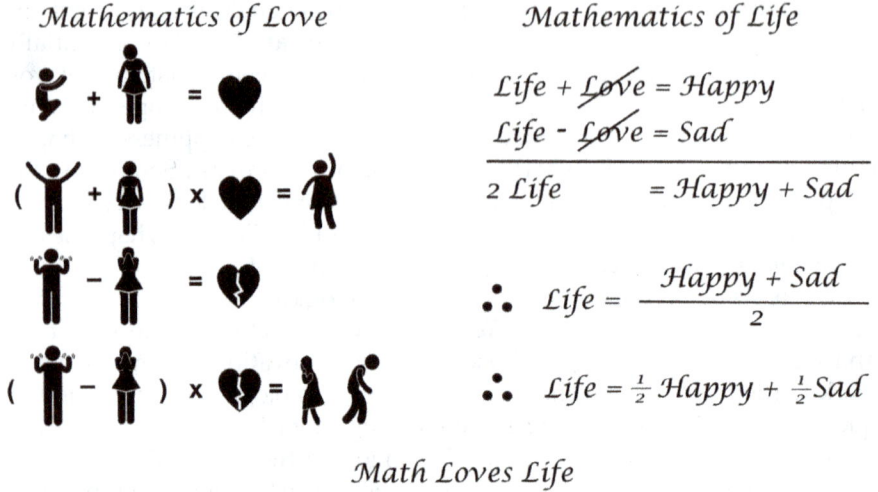

Mathematics of Love   *Mathematics of Life*

$$Life + Love = Happy$$
$$Life - Love = Sad$$
$$2\ Life \qquad = Happy + Sad$$

$$\therefore \quad Life = \frac{Happy + Sad}{2}$$

$$\therefore \quad Life = \tfrac{1}{2} Happy + \tfrac{1}{2} Sad$$

Math Loves Life

FIGURE 1.2
A metaphorical parallel between math and life.

empowerment with collective well-being. Only by addressing these complexities can we truly align our choices with happiness. As a form of entertainment, we draw an analogy or metaphor between life and mathematics, as illustrated in Figure 1.2.

In this analogy, we preserve the essential mathematical structure while substituting life-related terms. By incorporating symbolic language to evoke meaning and emotion, it also functions as a metaphor.

1.3 The Power of Analogies

Analogies are fundamental to human cognition, a bridge between the known and the unknown. They allow us to use familiar concepts to explain new or complex ideas, fostering understanding and creativity. Whether in statistics, science, or AI, analogies have played a pivotal role in shaping innovations and offering insights that would otherwise remain elusive.

The fundamental principle behind analogies is the similarity principle: "Similar things will behave similarly and the more similar they are the more similar they behave" (Chang, 2014, 2023).

This chapter will explore the origins of analogical thinking, examine why it is so powerful, and provide examples of how analogies have revolutionized fields from statistical theory to AI. Our journey begins with an exploration of the cognitive mechanisms that underpin analogical reasoning.

An analogy is a comparison between two things that are alike in some way, often used to explain a concept or idea by highlighting its similarity to something more familiar. The goal is to make the unfamiliar clearer by relating it to something known. Analogies usually involve a more detailed and explicit comparison than metaphors, often explaining how multiple aspects of one thing are similar to another.

A constructive example could involve using a geometric analogy to illustrate mathematical induction as follows.

Mathematical induction is a powerful tool in deductive (not inductive) reasoning. It is a proof technique commonly used to demonstrate that a given statement holds for all (or some) natural numbers. The method proceeds in two main steps:

1. Base case: For a statement that depends on a natural number n, establish that it holds true for an initial value n_0.

2. Inductive step: Assume the statement is true for a given natural number $n \geq n_0$. Then, show that if it holds for n, it also holds for $n+1$.

As an example, prove the following identity holds for any positive integer n:

$$1+3+\cdots+(2n-1)=n^2.$$

Here is the proof:

- The basis: It is obvious that for $n = 1$, the above equation holds.
- The inductive step: if the above equation holds for n, we prove that it holds for $n+1$. In fact,

$$\left[(1+3+\cdots+2n-1)\right]+\left[2(n+1)-1\right]=n^2+(2n+1)=(n+1)^2.$$

If we use geometric analogies, the proof may be illustrated using the diagrams in Figure 1.3.

At its core, an analogy involves comparing two different domains, identifying shared structures, and drawing insights from the relationship. For example, comparing an electrical circuit to a water system is a classic analogy that simplifies understanding of how current flows. But why do analogies work so well?

From cognitive science, it is shown that analogies tap into the brain's pattern recognition capabilities. They help bypass complex information processing by mapping a new situation onto a well-understood framework. This cognitive shortcut enables faster problem-solving and fosters creative leaps.

To harness the full potential of analogies in creative thinking and problem-solving, we often follow a systematic approach:

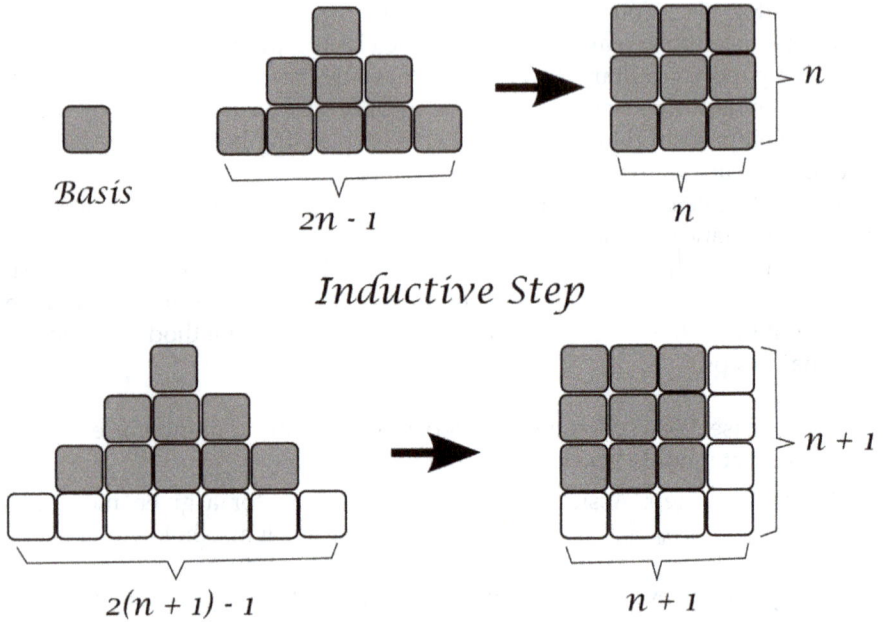

Basis

Inductive Step

$2n - 1$

n

Inductive Step

$2(n + 1) - 1$

$n + 1$

$n + 1$

Mathematical Induction

FIGURE 1.3
Geometric analogy explaining mathematical induction.

Analogy often starts from the problem to be solved. Then search for a source domain, looking for a domain that shares underlying structures, processes, or patterns with the target domain. The key is to find something that offers useful insights or analogies.

Once the source domain is identified, map the correspondences between the source and the target:

- Structural correspondence: A successful analogy must align the deep structures of both domains.
- Functional relevance: An analogy should highlight functional similarities that are useful for the problem at hand.
- Creative tension: The most creative analogies often introduce a level of tension between the two domains. This "creative tension" forces the mind to grapple with the differences between the two domains, often sparking innovative ideas.

To craft an innovative parallel, start by identifying the core principles that govern both the familiar and unfamiliar domains. Abstraction involves stripping away the details of both domains to focus on their core structures.

By abstracting the systems or processes involved, you can better identify deep similarities that are not immediately apparent. Thought experiments are often helpful during the process.

Analogical thinking often is an iterative process. After mapping the correspondences, test the analogy's validity and usefulness. By continually refining analogies and applying them to different aspects of the problem, you increase the likelihood of arriving at a novel solution.

An analogy does not always need to start with the problem itself; sometimes, it begins with a familiar principle or method, which is then applied to a new domain to spark unexpected insights or discoveries. Thus, fostering creativity involves actively keeping both intriguing questions and effective methods in mind, as the connection or revelation may come suddenly and unexpectedly.

For example, in Chapter 2, we delve into intransitivity in game theory, exploring intriguing phenomena like the counterintuitive outcomes of intransitive dice. Through analogy, we uncover that many of our everyday decisions—particularly those influenced by statistical theory in medical contexts—may be mere illusions. Building on this, Chapter 4 draws an analogy to the play-the-winner game to propose an innovative anti-polarization democratic system. For further illustration, in Chapter 14, we employ hierarchical analogies across different levels to reveal a profound insight: in logic, devolution and evolution must be equally probable. Data visualization serves as a unique analogy that connects data structures to visual elements, as discussed in Chapter 11.

1.4 Analogies in Engineering

For instance, the design of the Shinkansen bullet train in Japan was inspired by the beak of the kingfisher bird. Engineers observed that the kingfisher's beak allows it to dive into water with minimal splash, which led them to redesign the train's nose to reduce noise and drag when exiting tunnels at high speeds.

Similarly, the development of Velcro was inspired by the way burrs stick to animal fur. The analogy between burrs' tiny hooks and fasteners led to the creation of a new material that revolutionized fastening systems across industries, from clothing to space exploration.

Biomimicry is a perfect example of parallel thinking applied to innovation. In biomimicry, designers and engineers draw on analogies from nature to solve human problems. By studying how animals, plants, and ecosystems function, innovators have developed new materials, products, and systems that mimic natural processes.

One notable example is the development of synthetic materials inspired by spider silk. Spider silk is stronger than steel and more flexible than most synthetic fibers. Engineers studying the molecular structure of silk proteins were able to develop biomimetic fibers that replicate these properties, leading to innovations in fields ranging from medical sutures to bulletproof fabrics. The parallel between spider silk and synthetic materials goes beyond a surface-level comparison, delving into the molecular mechanisms that give silk its unique properties.

In cross-industry innovation, analogies drawn from one industry can lead to breakthroughs in another. For example, the automotive industry borrowed from aviation design principles to improve the aerodynamics of cars. By using analogous thinking from aircraft design, engineers reduced drag and improved fuel efficiency in cars, leading to significant advancements in transportation.

By drawing a parallel between natural fractals and the layout of cities, urban planners have designed more sustainable, adaptable, and efficient urban spaces. For example, fractal-inspired designs can be seen in the organic, branching street layouts of cities like Paris and the spatial organization of natural parks. These designs not only improve the aesthetics of the city but also optimize traffic flow, reduce energy consumption, and enhance livability.

Analogies often originate in one field, such as engineering, and extend their influence across various domains. For instance, dimensional analysis is a common tool in engineering models in laboratories. In Chapter 9, we discuss the tool and how to effectively use this tool in biomedical modeling. In Chapter 12, we discuss how to use Brownian motion and random walk methods to solve engineering problems and improve clinical trial efficiency via adaptive design.

1.5 Analogies in Science

Albert Einstein's development of the theory of general relativity was heavily influenced by analogical thinking. In his famous thought experiment, Einstein imagined a person inside an elevator in space. Without looking outside, the person would not be able to tell whether they were being pulled by gravity or accelerating upward in the elevator. This analogy led Einstein to his insight that gravity and acceleration are equivalent, a core principle of general relativity.

One of the most famous examples of analogy-driven discovery is the elucidation of the structure of DNA. The most significant analogy that the discoverers (James Watson and Francis Crick) used was the thinking of complementary base pairing (adenine with thymine and guanine with cytosine) as a "key fitting into a lock." This analogy helped them recognize how hydrogen bonds could stabilize specific pairings, ensuring a regular structure with consistent dimensions. They drew from the existing biological and chemical principles, such as the symmetry of molecular interactions, to deduce the spiral nature of the structure. While their reasoning was not explicitly expressed as "a twisted ladder," their physical models mirrored the ladder-like framework (with bases as rungs and the sugar-phosphate backbones as rails).

The analogy of CRISPR gene-editing technology to molecular "scissors" has transformed the way we think about genetic engineering. By comparing CRISPR to scissors that can precisely cut and paste DNA, scientists have made the complex process of gene editing more understandable and accessible. This analogy also paved the way for creative applications of CRISPR technology, such as developing treatments for genetic disorders and improving agricultural crops.

Analogies in science are often about simplifying complex systems. From Isaac Newton's analogy between the falling apple and the moon's orbit to Richard Feynman's analogy of the behavior of subatomic particles as a "wave," these comparisons shape how we understand physical and abstract phenomena.

In healthcare, analogies with nature have led to the development of biomimetic technologies. For example, the design of prosthetic limbs has been informed by studying the movement of animals. Researchers have also drawn analogies between the immune system and security systems to develop new cybersecurity models that mimic how the body defends itself against pathogens.

In drug development, the analogy between biological pathways (e.g., gene regulatory, protein-protein interaction, metabolic, signal transduction, epigenetic regulatory, and disease-specific networks) and complex communication systems has helped scientists understand how cells transmit signals. By comparing cell signaling to a series of on-and-off switches or relay systems, researchers can better target treatments that interrupt disease processes at critical points.

The tree of life analogy continues to inspire evolutionary biologists, helping them conceptualize the relationships between species, the mechanisms of genetic change, and the history of life on Earth. It also serves as a powerful visual tool for organizing vast amounts of biological data, such as in phylogenetic trees used in genomics.

Even more intriguingly, we explore how hierarchical analogy reveals why both evolution and devolution are equally probable outcomes. Hierarchical analogy gives rise to powerful recursive functions and self-referential paradoxes. For instance, consider the political argument where the poor claim, "You are rich—how could you understand the concerns of the poor?" The rich may counter, "You're not wealthy—how can you know that the rich don't understand?" Each side can continue the recursion indefinitely. Through this process, we can examine which perspective holds and explore the limitations of classical logic, as discussed in Chapter 10.

The play-the-winner strategy, originally developed in game theory, has been effectively applied in clinical trials to create more efficient response-adaptive designs. This strategy also offers a novel approach in political science, inspiring a democratic system aimed at reducing polarization. In this proposed system, a party's governing duration is directly linked to the proportion of votes it receives. The mechanics of this political model, along with its potential impact on governance, will be explored in detail in Chapter 4.

In Chapters 15, 16, and 17, we employ critical thinking and creative analogies to deeply explore these topics. These discussions lay a foundational framework for the subsequent chapters: Chapters 18 and 19.

1.6 Analogical Thinking in Mathematics and Statistics

Random sampling can be used to approximate a finite integral. Since integration is related to finding the area under a curve, and the area of a circle is tied to the constant π, we can draw analogies between these concepts. Using random sampling, we can address problems involving integration, area, and constants like π.

We know that $\sqrt{2}$ is not a rational number, meaning it cannot be expressed as a fraction p/q, where p and q are integers. The relationship $p^2 \neq 2q^2$ always holds true. More generally, for $\sqrt{k}$, where k is a prime number, the equation $p^2 \neq kq^2$ applies. This can be interpreted in graph theory: an integer square can represent the degrees of a fully connected network. It follows that the sum of the degrees of two identical fully connected networks and the degree of another fully connected network will never be equal.

Similarly, since $\sqrt[3]{3}$ is not a rational number expressible as p/q, the relationship $p^3 \neq 3q^3$ holds. This implies that we cannot transform cubic shapes (with integer edge lengths) into three identical cubic shapes, each also having integer edge lengths.

Mark-recapture methods (MRM) have a long history, originally developed for studying fish and wildlife populations before being adapted for various other purposes. Their application to epidemiological problems came later, benefiting from advances in statistical methods and techniques from other fields.

The simplest capture-recapture model, known as the two-sample model, estimates the unknown size of a population. In this model, the first sample involves marking or tagging individuals, who are then returned to the population. A second sample is later taken, and the number of marked individuals in this sample (the "recaptures") is used to estimate the total population size.

For example, to estimate the number (N) of fish in a pond without draining it, we first capture and mark M fish and return them to the pond.

Assuming the marked and unmarked fish are fully mixed by the next day, we capture another sample of size n. In this sample, the proportion of marked fish is expected to be $p = m/n$, where m is the number of marked fish recaptured. Since this proportion should also equal M/N, we can estimate the total fish population in the pond as $N = Mn/m$.

This method, often called the Lincoln–Petersen method, assumes:

1. Only two statistically independent samples are taken.
2. Each individual has an equal probability of being captured.
3. The population is closed, with no deaths, births, or migration between samples.

MRM is widely used in ecology to estimate population sizes and vital rates such as survival, movement, and growth. In medicine, it has been applied to estimate the number of individuals requiring specific services, such as those infected with HIV. It is also useful in fields like software engineering to study error rates in source code, clinical trials, and databases—any scenario where the total population size is unknown.

A creative analogy can be drawn for measuring the volume of water, V, in a pond using a similar approach. Instead of marking fish, we pour a small volume v of a colored, nontoxic test liquid into the pond and allow it to mix evenly through diffusion. A water sample is then taken, and the concentration c of the test liquid is measured. Since c is expected to equal v/V, the total volume V can be estimated using $V = v/c$. This method assumes v is negligible compared to V. This example demonstrates the versatility and adaptability of MRM for solving diverse estimation problems.

Intransitive dice involve three or more dice that probabilistically form a circular winning relationship: A wins against B, B wins against C, and C wins against A, as studied in Game Theory. When applied to decision-making problems, such as determining the optimal treatment in clinical trials, intransitive dice lead to a revolutionary concept of "probabilistically preferable." This concept provides a powerful resolution to the issue of intransitivity in decision-making. For more details, see Chapter 2.

By analogy to the "Fair-Game" Paradox, we can make two ineffective drugs appear effective through optimal randomization, without physically or chemically modifying the drugs. This approach will be discussed in detail in Chapter 3. Analogy enables us to apply the same method to different problems. Examples are demonstrated throughout the chapters.

Chapter 5 explores drawing analogies between cooperative games and clinical development programs, offering insights into optimizing trial designs and shortening the drug development cycle. Chapter 7 focuses on applying critical thinking to identify and mitigate various biases in analysis and decision-making. Chapters 8, 9, and 12 delve into analogies in statistical modeling, addressing diverse challenges and methodologies. These include estimating drug effect sizes without unblinding treatment codes, reducing model complexity through dimensional analysis, designing adaptive clinical trials using concepts from Brownian motion, and employing random-walk Monte Carlo methods to solve differential equations.

1.7 Analogies in Artificial Intelligence

Analogies have been instrumental in shaping artificial intelligence research. One of the earliest analogies drawn was between the human brain and a computer. The brain–computer analogy has guided the development of algorithms, neural networks, and even the field of machine learning, where neurons in the brain are compared to nodes in a neural network. While this analogy has its limits, it remains foundational for understanding AI development. In the most recent deep learning model (GPT), self-attention is mimicking the human's attention mechanisms.

Analogies in swarm intelligence and evolutionary algorithms (EAs) help bridge concepts from biology to AI, shedding light on ways to solve complex problems through distributed, adaptive systems.

Swarm Behavior as an Analogy: Swarm intelligence draws from the collective behavior seen in natural systems like ant colonies, bird flocks, and bee swarms, which efficiently solve problems through decentralized decision-making. In AI, swarm intelligence applies this to create algorithms that solve optimization and search problems by mimicking how individual agents in a swarm interact and adapt based on simple rules and local information.

Artificial ant colony optimization (ACO): Inspired by real ants' behavior in finding optimal paths to resources, the ACO algorithm uses artificial ants that simulate pheromone trails to locate optimal solutions. The analogy here allows ACO to solve complex problems like the traveling salesman problem and network routing by distributing agents across possible paths, allowing the most promising solutions to "emerge" based on collective exploration and reinforcement (pheromone deposition), as shown in Figure 1.4.

Natural evolution as an analogy: EAs borrow from biological evolution principles—natural selection, mutation, and reproduction—to iteratively improve solutions to complex problems. Genetic algorithms (GAs) and genetic programming (GP) are two prominent forms of EAs that simulate evolution to solve optimization, search, and machine learning problems.

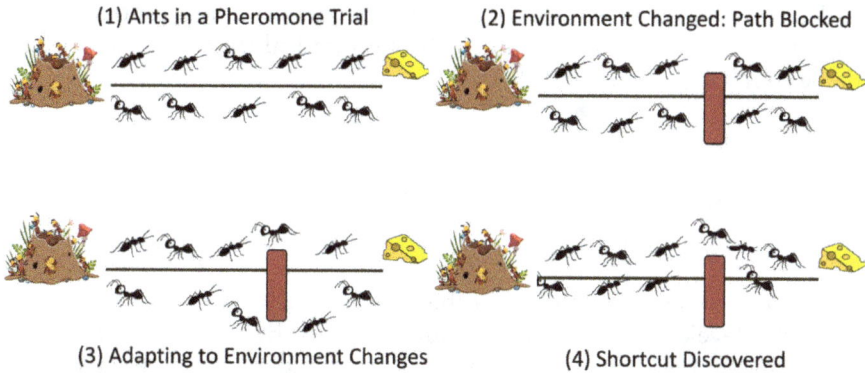

(1) Ants in a Pheromone Trial   (2) Environment Changed: Path Blocked

(3) Adapting to Environment Changes   (4) Shortcut Discovered

FIGURE 1.4
Swarm intelligence: adaptive behavior in ant colonies.

GAs: GAs simulate the "survival of the fittest" concept by representing potential solutions as a population of "individuals" with "chromosomes." These individuals undergo selection, crossover (recombination of parent traits), and mutation to explore the solution space. Analogous to evolution, only the best-performing individuals propagate to the next generation, gradually converging toward optimal solutions. This analogy allows GAs to tackle problems like machine learning model optimization, scheduling, and adaptive control.

GP: GP extends the GA concept to generate entire programs rather than specific solution values. It uses tree structures to represent potential solutions, evolving programs through crossover and mutation, which mirrors biological evolution but on functional programming structures. This approach has been applied in fields such as symbolic regression, automated design, and decision-making systems, where programs that maximize a fitness function are progressively refined.

Differential evolution (DE): DE is an optimization algorithm inspired by evolutionary principles, particularly suited for solving complex, non-linear, and non-differentiable optimization problems. It is part of the family of EAs and is widely used for global optimization tasks across various domains.

Through these analogies, swarm intelligence and evolutionary AI have emerged as powerful frameworks for exploring distributed intelligence and optimization, driving breakthroughs across diverse fields. Simultaneously, artificial neural networks powered by deep learning and big data have transformed nearly every aspect of life, spearheading advancements in healthcare, finance, and beyond. This synergy between nature-inspired algorithms and data-driven models underscores the vast potential of AI to tackle complex challenges and unlock new possibilities in science, technology, and daily life. These themes are further examined through critical thinking and creative analogies in Chapters 18 and 19.

1.8 Analogies in Design, Production, and Economics

In design thinking, the analogy of the "user journey" is frequently used to guide the development of products and services. By imagining the user's experience as a journey, designers can better understand how customers interact with their product over time, identifying pain points and opportunities for improvement.

This analogy allows designers to map out the entire lifecycle of a product from the user's perspective, leading to innovations that prioritize ease of use, satisfaction, and customer loyalty. The concept of a "journey" also encourages designers to think holistically, considering every aspect of the user experience.

One commonly used analogy in business is the SWOT analysis, which stands for *Strengths, Weaknesses, Opportunities*, and *Threats*. This framework is an analogy for assessing the internal and external factors that affect an organization. By drawing on the metaphor of a battlefield (where strengths and weaknesses represent internal forces, and opportunities and threats represent external forces), the SWOT analysis helps companies strategically position themselves in competitive environments.

The SWOT framework offers a straightforward yet insightful model for businesses to analyze their competitive landscape, formulate strategies, and evaluate risks. Similarly, drawing an analogy between production lines and Braess's Traffic Paradox can help create work policies that not only enhance efficiency but also promote social equity.

Blockchain technology is often explained through the analogy of a digital ledger. This comparison helps people understand how blockchain provides a secure, transparent way to record transactions, much like an accountant's ledger records financial transactions. The analogy has been crucial in explaining the decentralized nature of blockchain and its potential to revolutionize industries beyond cryptocurrency, such as healthcare, logistics, and governance.

The rise of companies like Uber, Airbnb, and Lyft is often described through the analogy of the "sharing economy," a concept rooted in collaborative consumption. This analogy reframed the way people thought about ownership and access to goods and services, leading to a wave of innovations in how people travel, work, and share resources.

One example of creative tension can be found in the analogy between Darwinian evolution and economic competition. The idea of "survival of the fittest" in nature was applied to market dynamics, leading to economic theories of competition and adaptation. While evolution and economics operate in fundamentally different domains, the creative tension between biological survival and business strategy led to innovative insights into market behavior.

For instance, Braess's Paradox—the counterintuitive idea that adding a road can increase rather than reduce traffic—began in transportation engineering but has since found wide applications in diverse industries, as discussed in Chapter 6.

1.9 The Importance of Knowledge in Logical Reasoning

Logical reasoning is a critical tool for constructing and understanding analogies, evaluating claims, solving problems, and uncovering truths. However, its effectiveness is deeply tied to the knowledge and context available to the person employing it. Without sufficient knowledge, even sound reasoning can lead to incorrect conclusions. Conversely, informed reasoning grounded in relevant knowledge can transform seemingly insurmountable paradoxes into solvable challenges.

1.9.1 The Story of the Universal Solvent

The story of the universal solvent illustrates the interplay between logic and knowledge. A person claims to have discovered a solvent that can dissolve any material, earning skepticism from others, including Mr. Edison, who challenges the claim by asking, "If this solvent exists, what container would you use to store it?" At first glance, this question appears to disprove the claim, as it seems impossible to contain such a substance without it dissolving its container.

However, with deeper knowledge—particularly from a chemist's perspective—the challenge becomes less compelling. A chemist might propose several plausible solutions:

1. Storing in specific conditions: The solvent could be stored in a dark, cold environment to reduce its activity.
2. Chemical separation: The components of the solvent could be stored separately and only combined when needed.

These possibilities highlight that Edison's reasoning, while logical, overlooks critical scientific insights, demonstrating how incomplete knowledge can limit the validity of logical conclusions.

1.9.2 Key Insights on Knowledge and Logical Reasoning

1. Logic requires context: Logical reasoning operates on premises, and the quality of those premises determines the reliability of the conclusion. If the premises are incomplete or based on flawed assumptions, even impeccable logic will yield incorrect results. In the universal solvent story, the implicit assumption that containment is impossible under all conditions reveals a lack of scientific understanding.

2. Knowledge expands possibilities: Expertise and knowledge introduce additional variables that can reshape the reasoning process. A person unfamiliar with chemistry may see Edison's question as definitive, while a chemist recognizes opportunities for exploration and innovation. Knowledge transforms "unsolvable" problems into practical challenges.

3. Challenging assumptions: Knowledge enables critical examination of assumptions that underpin logical arguments. In the universal solvent case, the assumption that containment is universally impossible is not self-evident but contingent on conditions like temperature, material engineering, and chemical reactivity.

4. The role of interdisciplinary thinking: Logical reasoning benefits from diverse knowledge bases. Scientific breakthroughs often occur at the intersection of disciplines, where insights from one field inform solutions in another. In the universal solvent example, insights from physics, engineering, or materials science could contribute to a viable containment strategy.

1.9.3 Analogies for Logically Explaining Complexity

Analogies serve as powerful tools for explaining complex phenomena by linking abstract or unfamiliar concepts to relatable scenarios. For example, in **physics**, electric circuits are often compared to water flowing through pipes. Voltage is likened to water pressure, current to the flow rate, and resistance to a constriction in the pipe. Just as higher pressure pushes more water through a pipe, higher voltage drives electrons through a conductor, while resistance limits the flow. This analogy simplifies the understanding of how resistors affect current, making electric circuits more intuitive.

In cognitive science, the brain is frequently described as a computer. The mind functions like software, running on the hardware of neurons and synapses. This analogy parallels how computers process, store, and retrieve data with how the brain processes sensory input, stores memories, and makes decisions based on experience. By framing cognitive processes as "programs," this comparison has significantly influenced artificial intelligence research, inspiring algorithms modeled on neural activity.

In biology, the immune system is compared to a military defense system. White blood cells act as soldiers, protecting the body against pathogens like viruses and bacteria. Specialized immune cells, such as phagocytes (foot soldiers) and antibodies (intelligence officers), identify and neutralize invaders. This analogy simplifies the complexity of immune responses and provides a framework for understanding vaccines as "training exercises" that prepare the immune system for real threats.

Knowledge about thought experiments further bridges abstract concepts with reality, offering analogies that test theories or provoke deeper insights. For instance, Galileo's Leaning Tower of Pisa experiment illustrates gravitational acceleration by imagining two objects of different masses falling simultaneously. This analogy challenges Aristotelian physics, making the abstract principle of equal gravitational acceleration intuitive. Similarly, the Trolley Paradox presents an ethical dilemma through a scenario where a runaway trolley forces a choice between sacrificing one person to save five others (Figure 1.5). This analogy highlights the conflict between utilitarian and deontological ethics. The Ship of Theseus raises questions about identity by asking whether a ship remains the same if all its parts are replaced over time. This thought experiment serves as a framework for exploring continuity and identity in both metaphysical and practical contexts, such as biological regeneration or technological upgrades.

By connecting the abstract to the familiar, analogies and thought experiments simplify complex ideas, making them accessible and fostering deeper understanding across disciplines, as seen throughout the book and specifically in Chapter 13.

In conclusion, knowledge is the foundation upon which logical reasoning operates. While logic provides the structure to evaluate and connect ideas, knowledge fills in the necessary context to make those evaluations valid and meaningful. As the universal solvent story demonstrates, logical reasoning without adequate knowledge risks drawing premature or incorrect conclusions. By cultivating expertise and remaining open to new insights, we can refine our reasoning and better address the challenges of the world.

FIGURE 1.5
The Trolley Paradox: visualizing an ethical dilemma.

1.10 Creative Analogies as a Tool for Innovation

Innovative analogies are more than cognitive shortcuts; they are catalysts for innovation. Many breakthroughs in technology, methodology, and ideas arise from recognizing parallels across distinct domains. Consider the famous example of the Wright brothers, who used analogies from bicycle mechanics to address the challenges of powered flight. By understanding that balance and control in flight could be approached similarly to steering a bicycle, they laid foundational principles for modern aviation.

This cross-pollination of ideas across disciplines enables radical breakthroughs. When we step back from rigid boundaries and use analogies to draw unexpected connections, we unlock creative solutions that transcend conventional thinking. For instance, in Chapter 17, we use recursive analogies structured around "explained by" to explore understanding itself. This leads to a concept-mapping framework where knowledge is organized in a circular, hierarchical network—making it possible to apply network analysis tools to analyze the meaning of concepts regardless of specific natural language used.

In Chapters 2, 3, and 5, we explore the efficient clinical trial design and beyond through analogous randomizations.

By combining various analogical tools, we ultimately examine the philosophical foundations, architectures, and prototypes of fully humanized AI, integrating constructivist and behaviorist approaches (see Chapter 19).

Cultivating analogical thinking requires openness to new perspectives and a willingness to venture into diverse fields. Here are some strategies to develop analogy as a powerful tool for creative thinking:

1.10.1 Embrace Curiosity and Challenge Assumptions

An analogical mindset begins with curiosity. By staying curious about the world around us, we remain open to discovering new connections between ideas, systems, and experiences. Whether in science, business, or daily life, curiosity drives us to ask questions and seek patterns that lead to analogical insights. By nurturing a sense of wonder and exploration, we create the mental conditions for analogies to emerge.

Effective analogical thinking often involves challenging our assumptions and seeing the familiar in a new light. This requires a willingness to question established norms and consider alternative perspectives. By embracing ambiguity and complexity, we open ourselves to analogies that defy conventional logic and lead to novel insights.

1.10.2 Broaden Knowledge Base and Practice Cross-Disciplinary Thinking

To draw effective analogies, you need a wide range of knowledge and experiences to draw from. Exposure to multiple disciplines, whether through reading, travel, or collaboration, is required to enrich your mental reservoir of analogical possibilities. The more diverse your knowledge base, the more potential analogies you can create when solving problems. Cross-disciplinary thinking fosters innovation by allowing us to transfer knowledge across domains and apply it in new and unexpected ways.

1.10.3 Look for Structural Similarities

When seeking an analogy, look for structural similarities between domains, rather than focusing on surface details. For example, the structure of a neural network in AI and the structure of biological neurons share deep similarities, even though one is artificial and the other biological. The power of analogy lies in recognizing these deep, often hidden, parallels that can lead to creative insights.

1.10.4 Discover General Rules via Analogy

The identification of common structures based on the Similarity Principle across diverse fields forms the basis of scientific laws. These laws or theories capture patterns, relationships, or behaviors observed in nature or systems. The process of recognizing these underlying similarities, especially when drawn from specific examples or limited data, is known as induction. Through abstract thinking, these analogies and common structures are further refined into mathematical models. Abstract reasoning allows for the mathematical representation of the principles, distilling complex phenomena into generalized equations or models. This process often transforms intuitive analogies into formal scientific laws, enabling broader application and predictive power across disciplines.

1.10.5 Thought Experiment: Analogy in Mind

A thought experiment is a hypothetical scenario created in the mind to mimic an imaginary situation and to explore the consequences, implications, or principles of a concept, theory, or idea without the need for physical experimentation. Thought experiments are used extensively in philosophy, physics, mathematics, and other fields to reason about complex issues, test hypotheses, or challenge existing beliefs.

Thought experiments, much like those used by Einstein, are powerful tools for developing analogies. By imagining hypothetical scenarios or placing yourself in a different context, you can explore new ways of thinking about a problem. Thought experiments free the mind from practical constraints, allowing it to explore creative analogies without the burden of real-world limitations.

For example, the Trolley Problem is an ethical dilemma: There is a runaway trolley barreling down a stretch of railway tracks. Ahead, on the tracks, there are five people tied up and unable to move. The trolley is headed straight for them. You are standing some distance off in the train yard, next to a lever (Figure 1.5). If you pull this lever, the trolley will switch to a different set of tracks. However, you notice that there is one person on the sidetrack. You have two (and only two) options:

1. Do nothing, in which case the trolley will kill the five people on the main track.
2. Pull the lever, diverting the trolley onto the side track where it will kill one person.

Which is the more ethical option?

1.11 Analogies as Artful Innovation and a Journey Worth Taking

Analogical thinking is a journey of exploration, connecting distant ideas in meaningful ways. As we have seen, analogies serve not only as educational and problem-solving tools but also as a driving force behind some of the most important discoveries and inventions in history. From Watson and Crick's DNA double helix to the development of AI neural networks, analogies have fueled humanity's quest for understanding.

The art of analogy is central to creative thinking and innovation. Whether in science, technology, or engineering, analogical thinking has the power to unlock new perspectives, enabling breakthroughs that push the boundaries of what is possible. By recognizing deep structural similarities between different domains, we can leverage analogies to solve complex problems and drive radical innovation.

In the following chapters, we will delve into the crucial role of analogies across diverse fields, investigating how they have deepened our understanding of statistics, science, and artificial intelligence while fostering creativity and innovation. We will explore how paradoxes serve as a guiding theme, using analogies to bridge disparate disciplines and uncover unexpected connections. These challenges will be addressed through a blend of mathematics, statistics, computer simulations, thought experiments (analogous experiments in mind), and logical analysis, unveiling novel insights and potential solutions.

Our emphasis is on fostering conceptual understanding, critical thinking, and creative analogy. As such, most chapters focus on ideas and applications rather than heavy mathematics, though concepts such as Brownian motion, threshold regression, and partial differential equations are lightly explored. However, in a few chapters, we provide full statistical proofs to illustrate how rigorous critical thinking and creative analogy can be applied to generate meaningful and practical outcomes.

As the book progresses into later chapters, the depth and complexity of critical thinking and creative analogies grow. This evolution culminates in discussions on topics like *Analogies in Artificial Intelligence* and *Humanized AI*, where the interplay between analogy and innovation becomes both intricate and profound. I hope you find the book both interesting and inspiring, and I look forward to meeting you in the final chapter: *Education Shifts and Futuristic Society in the AI Age*.

1.12 About Exercises

Exercises are provided for each chapter, covering five key aspects: *Understanding Key Concepts, Critical Thinking and Application, Analogy and Creativity, Debate and Discussion,* and *Exploring Extensions*. The key concepts in each chapter are clearly identified, encouraging students to explore connections through analogies or similarity searches. The outcomes of these exercises may include short essays accompanied by diagrams to illustrate insights. Furthermore, these exercises are designed to foster connections across chapters, promoting an integrated understanding of the material.

Students are encouraged to break free from traditional constraints and not limit their imagination to the information presented in the chapter. Emphasis is placed on critical thinking and creative analogies, while mathematical derivations—though included in a few chapters—are de-emphasized due to the availability of AI tools. The utilization of AI tools is not prohibited but is typically recommended only after students have manually worked through the problems. This approach reflects real-world scenarios, where AI serves as a supportive tool rather than a primary resource, enhancing both understanding and problem-solving skills.

The AI age is defined by the value of questions over answers, adaptability over stability, and connections over isolated knowledge. Therefore, readers and students are encouraged to discover problems to solve through the use of analogies, fostering creativity and deeper engagement with the material.

In addition to the chapter abstracts, key concepts (terms) are highlighted in colored text. Connecting these terms through lines of analogy or similarity search, both within and across chapters, is an effective way to deepen understanding and learn analogical thinking.

Exercise

Understanding Key Concepts

1. Define critical thinking and creative analogy. Provide examples of how each can be applied to solve problems in different fields, such as science or business.

2. Discuss the Similarity Principle and its role in analogy, creativity, and problem-solving. How does it guide reasoning and innovation?

Critical Thinking and Application

3. Explain how critical thinking and creative analogy complement each other in problem-solving. Use an example, such as optimizing traffic systems or explaining ecosystems.

4. Identify potential pitfalls when using analogies for problem-solving. How can critical thinking mitigate these challenges?

Analogy and Creativity

5. Develop an analogy to explain a complex concept (e.g., machine learning and the immune system). Highlight how this analogy simplifies understanding while retaining accuracy.

6. Identify a historical or modern example where an analogy led to a major breakthrough (e.g., Velcro and DNA structure). Explain how the analogy influenced the solution.

Debate and Discussion

7. Debate whether focusing too much on logic (critical thinking) can hinder creativity. How can individuals balance structured reasoning with imaginative exploration?

8. Discuss how analogies can be used to navigate ethical dilemmas (e.g., the Trolley Problem). What are the strengths and limitations of using analogies in ethical reasoning?

Exploring Extensions

9. Research an analogy from a different discipline (e.g., the brain as a computer in AI and the immune system in cybersecurity). Explain how this analogy provides insights and facilitates understanding.

10. Propose a framework to systematically develop and evaluate analogies for teaching or problem-solving. Include criteria for identifying useful analogies and ensuring their validity.

11. Comedians are particularly known for their sharp use of analogies to craft insightful and hilarious observations. But why and how do they excel in this?

2

The Golden Point for Decision-Making

The true golden point for decision-making is the mathematical golden point, where balance, proportion, and harmony converge to yield the most optimal and elegant solution.

2.1 Paradox of Winning

In a world that often craves order and predictability, intransitivity serves as a fascinating reminder of life's inherent complexity. It challenges our instinct for simple hierarchies, revealing dynamics that defy straightforward rankings.

In nature, the side-blotched lizard provides a vivid example. Male lizards exhibit three distinct mating strategies—orange, blue, and yellow—each cyclically outcompeting the other in a rock-paper-scissors-like pattern. This intransitivity shapes population dynamics in ways that are both intricate and unpredictable.

In sports, intransitivity appears when Team A beats Team B, Team B beats Team C, and Team C beats Team A, forming cyclic victories instead of a clear hierarchy. Similar patterns are observed in warfare and economics, where context and strategy often determine outcomes, not absolute superiority.

Intransitivity forces us to rethink linear rankings, encouraging adaptive, context-sensitive approaches over rigid frameworks. A playful way to explore this concept is through intransitive dice, where each die is designed to "win" against one opponent but "lose" to another in a continuous cycle. Intransitive dice not only illustrate the phenomenon visually but also allow us to study intransitivity statistically, providing a bridge to a deeper mathematical exploration.

Intransitive (nontransitive) dice are familiar to us in game theory. However, when they are introduced in decision problems without an opponent, they reveal some profound impacts that can leave us unsettled. Suppose we have a set of three intransitive dice: A has sides $\{2, 2, 4, 4, 9, 9\}$, B has sides $\{1, 1, 6, 6, 8, 8\}$, and C has sides $\{3, 3, 5, 5, 7, 7\}$, as shown in Figure 2.1. When rolling the dice and determining a higher value as the winner, A will win over B, B

DOI: 10.1201/9781003630081-2

Intransitive Dice

A   B   C

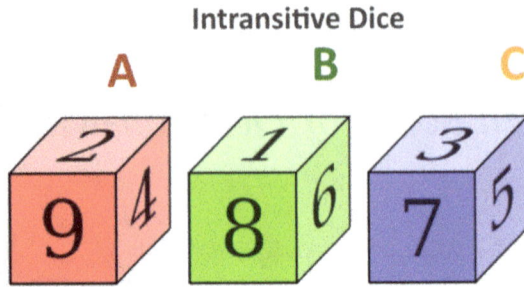

- *A* has sides 2, 2, 4, 4, 9, 9.
- *B* has sides 1, 1, 6, 6, 8, 8.
- *C* has sides 3, 3, 5, 5, 7, 7.

A > *B* > *C* > *A*, with 56% probability.

FIGURE 2.1
Three intransitive dice: an illustration of non-transitive properties.

will win over *C*, and *C* will win over *A*, each with a probability of 56%. This creates a circular winning situation.

Intransitivity is not always desirable. Imagine dice *A*, *B*, and *C* represent three different drugs or medical interventions, with their face values corresponding to potential responses of patients who may take the respective drugs (here, we implicitly draw an analogy between dice and medical treatments). In clinical trials, if there are only two randomized groups with responses *A* and *B*, we may conclude that *A* is better than *B*, at least temporarily, without further analysis—even though they may actually be equally effective. Since many non-parametric hypothesis test procedures, such as one-sided Wilcoxon tests, sign test, permutation test, win-ratio tests, and the log-rank test, rely on this notion of winning, it is essential to closely examine their validity in practical applications, even if they are statistically valid in pairwise comparisons.

It is evident that even if *A* wins against both *B* and *C*, transitivity is not guaranteed. There could exist a die *D* such that *A* wins against *B*, *B* wins against *D*, and *D* wins against *A*. Additionally, the discrete values on dice can represent more faces or even probability distributions without altering the intransitivity.

When dealing with intransitive problems, it is crucial to distinguish between game theory and classic decision theory. In game theory, the choice you make often depends on your opponent's choices and the available options. Conversely, in the classical decision theory, the ranking between any two choices should depend solely on the individual rankings of those two choices, independent of other options.

The concept of intransitivity with probability distributions has applications in various fields. In social choice theory, it examines voting systems where different procedures might favor distinct candidates depending on the distribution of voter preferences. Similarly, in the election theory, intransitive preferences in voting systems can be compared to intransitive dice. For example, in scenarios with three or more options (candidates, policies, etc.), a majority may prefer candidate A over B, B over C, and yet C over A, illustrating the Condorcet paradox.

2.2 Practical Intransitive Situations

Chang (2019 and 2023) briefly explored **intransitive dice** in the context of medical treatment selection and the controversies surrounding one-sided rank test procedures, but these discussions lacked detailed analysis. We can demonstrate and rigorously prove that, whether stratification is applied or not, the intransitivity of treatment effects on winning probabilities persists. This persistence arises from the fundamentally ill-defined nature of preferences or winning criteria in such contexts. Unlike game theory, classical decision-making problems—such as treatment selection in clinical trials—do not involve a clear player-opponent framework. Instead, our focus is broader and more fundamental, addressing the intransitivity inherent in decision-making processes tied to various one-sided nonparametric test procedures.

In clinical trials, patients often belong to subgroups defined by demographic factors and baseline characteristics, leading to complex response distributions to medical interventions. For instance, when subgroup responses follow distinct distributions (sub-distributions), the overall population response forms a mixture of these sub-distributions. Specifically, if responses in the test and control groups follow two distinct normal distributions, the overall population response will result in a mixture of these normal distributions. This complexity underscores the challenges of drawing meaningful conclusions in treatment selection and highlights the role of intransitivity in such analyses.

You might wonder whether having two random variables that are either single-parameter distributions or unimodal distributions ensures transitivity. First, a single-parameter distribution is not necessarily unimodal. Second, two unimodal distributions do not guarantee transitivity in comparison. Here is an example illustrating X, Y, and Z forming intransitive winning probabilities: X and Y are normal distributions $N(1,1)$ and $N(0,5^2)$, respectively, while Z is a normal mixture of $N(-7.3, 0.4^2)$ with a proportion of $1/3$ and $N(2.4, 0.4^2)$ with a proportion of $2/3$ (Figure 2.2). The cumulative

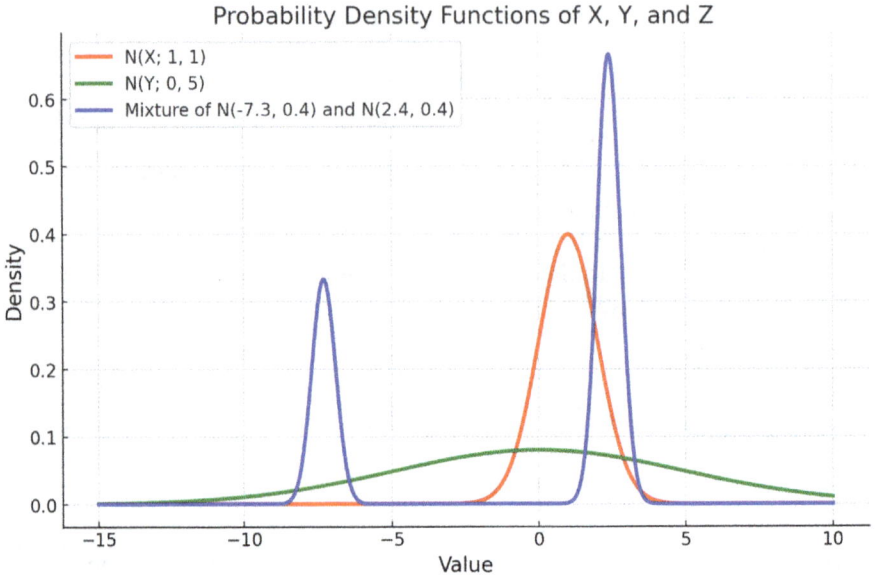

FIGURE 2.2
Probability density functions of normal and mixture distributions.

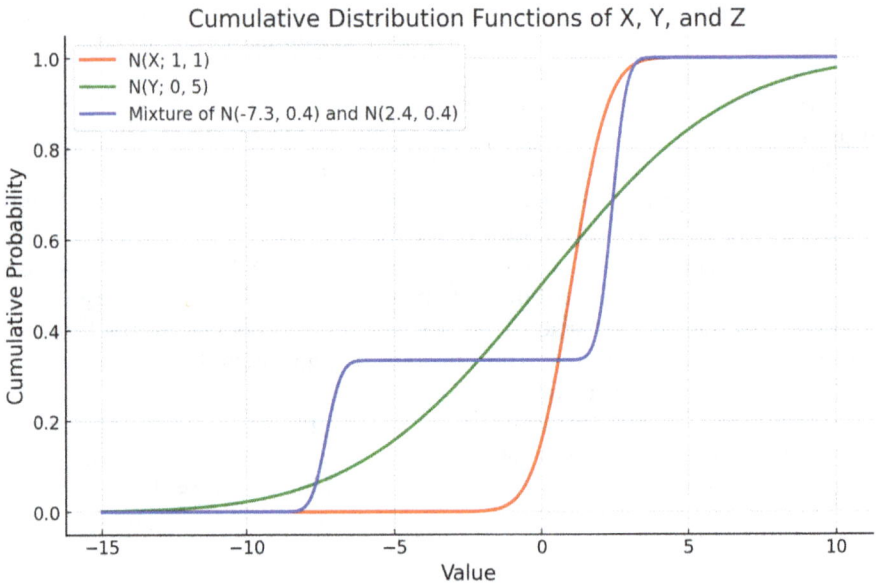

FIGURE 2.3
Cumulative distribution functions of normal and mixture distributions.

distribution functions (CDFs) are shown in Figure 2.3, and the winning probabilities are $P(X > Y) \approx 0.577$, $P(Y > Z) \approx 0.520$ and $P(Y > Z) \approx 0.602$, resulting in an intransitive situation.

2.3 A Sufficient Condition for Transitivity

Identifying a sufficient condition to avoid intransitivity and ensure transitivity is both interesting and important. We can prove that for two independent random variables X and Y, complete stochastic dominance (CSD) of X over Y is a sufficient condition for transitive winning probabilities (Chang and Lu, 2024). CSD occurs when the CDFs of two variables do not intersect at any point. However, CSD is not a necessary condition. Transitivity can still hold under certain distributional constraints without requiring CSD, as discussed below.

CSD might indeed be a stringent requirement. By imposing specific distributional assumptions, such as the proportional hazard assumption in survival analysis, we can often satisfy the transitivity criterion more readily. For example, Lebedev (2019) demonstrated that intransitivity does not occur for many classical continuous distributions, including uniform, normal, logistic, and Laplace distributions with equal variances, as well as exponential and Weibull distributions when one parameter (mean or variance) is equal.

A more general conclusion can be drawn: for any single-parameter distributions (with a location parameter c) of X and Y, intransitivity is impossible as long as X and Y are independent. In such cases, $\text{CDF}(X) = \text{CDF}(Y + c)$, and due to the monotonicity of the CDF, CSD is guaranteed. These assumptions are commonly applied in clinical trials. However, if such strong distributional assumptions are imposed, the benefits of using nonparametric tests may be diminished, as discussed later in this paper.

In clinical trials, making distributional assumptions about X and Y might be justifiable. However, extending these assumptions to a potential third variable Z for any treatment is more complex. The derived sufficient conditions for transitivity under such restrictions may be difficult to interpret and challenging to apply in practice.

2.4 Intransitivity Issue in Hypothesis Testing

Nonparametric tests such as the Wilcoxon Test, the Sign Test, the Wilcoxon Rank Sum Test, the Log-Rank Test, and the more recent Win-Ratio Test are widely used in clinical trials and other scientific experiments. These tests offer significant advantages over parametric counterparts, particularly when working with non-normal distributions, ordinal or nominal data, small sample sizes, interval or ratio data, or datasets containing outliers. Their flexibility and minimal assumptions make them invaluable tools in diverse research contexts.

Clinical trials are structured studies designed to evaluate the safety, efficacy, and potential side effects of medical interventions, such as drugs, therapies, or diagnostic methods, under controlled conditions. Hypothesis testing plays a central role in these studies, providing a statistical framework for assessing the effectiveness of interventions. Typically, the null hypothesis asserts no difference between a new treatment and a control (e.g., a placebo or standard therapy), while the alternative hypothesis suggests a positive effect. Researchers analyze data collected during the trial's various phases using statistical tests to determine whether sufficient evidence exists to reject the null hypothesis, thereby supporting the intervention's efficacy. Rigorous hypothesis testing ensures that conclusions are reliable, reducing bias and enhancing the credibility of clinical trial results.

However, nonparametric tests, particularly in clinical trials, often face challenges in reliably identifying the superior treatment. A key limitation arises from the issue of intransitivity, discussed earlier. This issue is analogous to the concept of winning probabilities in dice games, where non-transitive and circular winning scenarios occur naturally, even before any specific test procedure is applied. Intransitivity is not confined to rank-based test methods; it also extends to nonparametric approaches reliant on sample medians.

Intransitivity also exists in correlated data. For instance, in Stratum 1, the responses are (2, 1, 3); in Stratum 2, they are (4, 6, 5); and in Stratum 3, they are (9, 8, 7) for drugs A, B, and C, respectively. Here, the probabilities of one treatment being superior to another exhibit circular patterns: $P(A>B)=P(B>C)=P(C>A)=2/3$. This highlights the inherent challenge of intransitivity in certain nonparametric testing scenarios.

2.5 The Golden Point: Resolution to Intransitivity Problems

CSD is a strong requirement that is often difficult to meet in practice. To find a practical resolution, when we make a decision facing multiple (e.g., 3) options, we are not only comparing which probability is bigger but also how much bigger. From this idea, we introduce a stronger form of preference, "Probabilistically Preferable ($\succ$))," defined as:

$$A \succ B, \text{ iff } P(A > B) > \max(\min(P(B > C), P(C > A))),$$

where maximum is taken over all possible random variables C.

Probabilistically preferable (PP) says in a layman's language that if and only if $P(A > B)$ is not the smallest of the three probabilities for all possible C, we conclude $A \succ B$, though we cannot conclude A is the best among A, B, and C.

Why is PP meaningful practically? This is because of the following:

First, PP not only compares which probability is larger but also considers the probabilistic magnitude of the difference to prevent intransitivity.

Second, when A is PP to B, B cannot win over (or get even with) A neither directly nor indirectly as proved below.

1. Under PP, let $C = B - \varepsilon$, where real value $\varepsilon > 0$,

$$\min(P(B > C), P(C > A)) = P(B - \varepsilon > A) \to P(B > A) \text{ as } \varepsilon \to 0.$$

   Thus from $A \succ B$ implies $P(A > B) > P(B > A)$.

2. For B to beat A, B has to beat C and C needs to beat A. The probability of B beating A via chain BCA is denoted by $P_{BCA}(B > A)$. Under PP, $A \succ B$, we have

$$P_{BCA}(B > A) \leq P(B > C)P(C > A) < \min(P(B > C), P(C > A)) < P(A > B).$$

This means B cannot beat A through the chain. In fact, B cannot beat A through any chain that includes B with n (>1) random variable because adding a variable in the existing chain will not increase the winning probability because of the probability multiplication rule and C is any random variable.

Last, PP will guarantee transitivity and avoid intransitivity completely as shown below.

The maxmin probability of cyclic random variables for each integer $n \geq 3$ is defined as

$$\pi_n = \max \min \left(P(U_2 > U_1), \ldots, P(U_n > U_{n-1}), P(U_1 > U_n) \right),$$

where the maximum is taken over all sets of independent random variables $U_1, \ldots, U_n$.

The intransitivity means meet $A \succ B$, $B \succ C$, and $C \succ A$, simultaneously. Thus to prove transitivity, we only need to prove the following inequalities cannot hold at the same time:

$$P(A > B) > \max(\min(P(B > C), P(C > A))), \tag{2.1}$$

$$P(B > C) > \max(\min(P(C > A), P(A > B))), \tag{2.2}$$

$$P(C > A) > \max(\min(P(A > B), P(B > C))). \tag{2.3}$$

For any given A, B, and C, at least one of $P(A > B)$, $P(B > C)$, and $P(C > A)$ is the smallest among the three (two or all three may be equally small). However, if $P(A > B)$ is the smallest, inequality (2.1) is false; if $P(B > C)$ is the smallest, inequality (2.2) is false; if $P(C > A)$ is the smallest, inequality (2.3) is false. That is, at least one of the inequalities is false. This proves the new definition of preference is always transitive.

The concern is that PP can avoid intransitivity problems, but is difficult to use. How can we develop a practically simple way to ensure PP based on $P(A > B)$ alone?

To derive a sufficient condition in terms of value of $P(A > B)$, we use the result for the maxmin probability π_n (Nagy, 2011; Vuksanovic and Hildebrand, 2021),

$$\pi_n = 1 - \frac{1}{4\cos^2\left(\pi/(n+2)\right)}, \, n \geq 3.$$

For *n*=3, it reduces to

$$\pi_3 = 1 - \frac{1}{4\cos^2\left(\pi/5\right)} = \frac{\sqrt{5}-1}{2} = 0.618.$$

Interestingly, 0.618 is the well-known golden point. Therefore, if $P(A > B) > 0.618$, we know $P(A > B)$ is not the smallest among the three probabilities $P(A > B)$, $P(B > C)$, and $P(C > A)$ for any C, meaning $P(A > B) > \max(\min(P(B > C), P(C > A))$ for any C. Thus, we can conclude $A \succ B$ or treatment A is better than treatment B. The true golden point for decision-making is the mathematical golden point, where no ambiguity is allowed.

However, the sufficient condition $P(A > B) > 0.618$ is not a necessary condition for transitivity because, for example, given normal distributions $A \sim N(0.1, 1)$ and $B \sim N(0, 1)$, the probability $P(A > B) = 0.5283 < 0.618$, but A and B will not cause intransitivity for any C.

This result for independent A, B, and C suggests that given medical treatment or other decision options A and B, $P(A > B) > 0.618$ or $P(A > B)/(P(B > A)) > 1.618$ (conjugate golden ratio) is a sufficient condition for transitivity, which concludes option A is better than option B in the PP sense. This sufficient condition is more feasible than CSD in most practical situations. To ensure transitivity among any three or more variables, set n→∞ in the π-formulation. This yields a sufficient condition for transitivity: $P(A > B) > 0.75$.

2.6 Conclusion: Understanding and Addressing Intransitivity

Intransitivity is a pervasive phenomenon, commonly observed and accepted in games, sports, biological evolution, and warfare. However, it poses significant challenges in critical contexts such as medical and daily decision-making, where consistency and reliability are essential.

The key takeaways from this chapter flow logically as follows:

1. Intransitivity of winning is as prevalent in continuous outcomes as in discrete outcomes. It can arise even in simple distributional forms. Many decisions we consider "rational" may be illusions if we fail to recognize the role of intransitivity.

2. Intransitivity is not influenced by sample size or p-value but stems from the inherent ambiguity in defining preferences based on winning probability.

3. The commonly used one-sided nonparametric rank tests often fail to account for intransitivity in clinical trial decision-making.

4. A PP framework effectively addresses intransitivity by considering the magnitude of probabilities.

5. The role of the golden point (0.618) serves as a practical transitivity rule: If the probability of preferring A over B exceeds 61.8% $(P(A>B)>0.618)$ for three variables or 75% for any number of variables, transitivity is maintained. This provides a simple yet powerful guideline for decision-making.

This exploration of intransitivity provides broader lessons in critical thinking and creative analogy, aligning with the central theme of this book:

6. Recognizing patterns and ambiguities: Critical thinking requires identifying complex patterns and potential ambiguities, such as the intransitive relationships inherent in decision-making frameworks.

7. Challenging assumptions: Creative analogy involves questioning standard assumptions, such as the belief that nonparametric statistical tests can inherently be used to identify better treatment, and proposing alternative frameworks like probabilistically preferable criteria.

8. Integrating insights across domains: By drawing analogies from game theory, probability, and clinical decision-making, this chapter demonstrates the value of interdisciplinary thinking in uncovering innovative solutions.

9. Simplifying complexity: The golden point (0.618) exemplifies the power of distilling complex concepts into actionable guidelines, making them accessible and practical for decision-makers.

10. Avoiding cognitive illusions: Both critical and creative thinking involve recognizing and avoiding cognitive traps, such as the illusion of transitivity in everyday decisions, ensuring that energy and resources are directed toward meaningful efforts.

By combining rigorous analysis with creative analogy, this chapter not only addresses a specific issue—handling intransitivity in clinical decision-making—but also provides a framework for thinking critically and innovatively about broader challenges. These principles will continue to guide discussions in subsequent chapters, fostering a deeper understanding of the interplay between logic, creativity, and practical decision-making.

Exercise

Understanding Key Concepts

1. Explain the concept of intransitive dice and how they create a circular preference. Use an example from the chapter, such as the dice with face values {A, B, C}.

2. Define the concept of "Probabilistically Preferable." Explain how it ensures transitivity in decision-making and why it is a robust alternative to other methods.

3. What critical thinking ideas and elements of creative analogies stand out to you in this chapter?

Critical Thinking and Application

4. Discuss how intransitivity issues in statistical decision-making could impact clinical trials. Provide an example of how using rank tests might lead to unintended interpretations.

5. Explain why the golden point (0.618) is significant in resolving intransitivity problems. How does it provide a sufficient condition for transitivity?

Analogy and Creativity

6. Develop an analogy to explain the concept of "Probabilistically Preferable" to someone unfamiliar with statistics. Use a real-world decision-making scenario, such as choosing a meal at a restaurant.

7. Identify a real-world situation (e.g., sports rankings and election results) where intransitivity occurs. Create an analogy to explain this phenomenon to a general audience.

Debate and Discussion

8. Debate whether it is ethical to base medical treatment decisions on statistical tests prone to intransitivity. Should more robust frameworks like PP be mandated in clinical trials?

9. Discuss how to use hypothesis test methods to test the null hypothesis: PP<0.618?

Exploring Extensions

10. Explore how the concept of the golden point (0.618) might be applied to other fields, such as economics or voting systems. Provide an example.

11. Research and propose an alternative framework or method that could address intransitivity issues in decision-making. Compare it to the PP framework.

12. Why is the golden point of transitivity applicable to any set of n dice ($n>2$) in terms of probabilistic preference? Provide a mathematical proof or a verbal explanation to support your argument.

3

The Art of Randomization

A game's fairness is shaped by its players.

3.1 The Paradox of Fair Game

Chang (2014) discusses a "fair game": Two opponents, A and B, play a game where each constructs a stack of one-dollar coins. To play, they simultaneously remove a coin from the top of their stacks. The coin cannot be flipped over. The rules for winning are defined as follows:

1. If both coins show heads (H, H), A wins $10 from B.
2. If both coins show tails (T, T), there is no winner, and A wins $0.
3. If one coin shows heads and the other tails (H, T or T, H), B wins $5 from A (Figure 3.1).

If player A randomly shows heads or tails with equal probability, then (1) player B always shows heads and (2) player B always shows tails; in both scenarios, B can expect to win zero per round over many rounds. Thus, one

Game Rules:
1) If heads-heads appears, Andy wins $10 from Bob.
2) If tails-tails appears, no winner.
3) If heads-tails or tails-heads appears, Bob wins $5 from Andy.

FIGURE 3.1
The paradox of a fair game: expected value mismatch.

DOI: 10.1201/9781003630081-3

might assume the game is fair if the average gains when A wins ($10/2) and B wins ($5/2+$5/2) are equal, suggesting that the arrangement of the coins does not matter if the opponent is unaware of the stack's order. However, this is not the case. For example, if B arranges his coins such that a quarter of them, randomly distributed, are heads, he can expect to win an average of $1.25 per round over many rounds. Here is why: Let a and b represent the fraction of their respective stacks that A and B arrange with heads facing up. The expected gains per round for player B are calculated as follows:

$$G = -10ab + 5a(1-b) + 5b(1-a).$$

Interestingly, if $b = 0.25$, then $G = \$1.25$ regardless of the value of a.

3.2 Analogies in Clinical Trial Randomization

Randomization is a cornerstone of clinical trial design, ensuring the unbiased assignment of participants to different treatment groups. By randomly allocating individuals to experimental or control arms, randomization is a reorganized approach in minimizing selection bias and balancing known and unknown confounding variables across groups. This creates comparable groups and strengthens the internal validity of the trial, making it possible to attribute differences in outcomes directly to the intervention being tested. Randomization also underpins the statistical framework for hypothesis testing, enabling valid inferences about the treatment's effectiveness. Various methods, such as simple randomization, block randomization, or stratified randomization, can be employed to achieve balance, depending on the study's design and objectives. Ultimately, randomization is essential for generating reliable, high-quality evidence to inform medical decision-making.

The intriguing "fair game" finds real-life applications, such as in medical treatments during clinical trials. For example, it is possible to make two individually ineffective drugs work effectively by using appropriate randomization strategies, without making any physical or chemical alterations to the drugs themselves. Here is how the analogy works:

3.2.1 Key Analogies

1. Player A has special coins with heads representing patients of biomarker H and tails representing patients of biomarker T (negative H). The overall patient population consists of H with proportion a and T with proportion $1-a$. He chooses patients of H or T randomly.

2. Player B has special coins with heads representing prescription drug H and tails representing drug T. He chose drug H or T randomly with randomization ratio $b : 1-b$.
3. Play B's winning amount G is analogous to the average patient benefit or effect of combined treatment (drug H or T), which depends on the treatment randomization ($a:1-a$) and population composition (b versus $1-b$).

Thus, the analogy to the fair game and the positive gain of Player B offers a subtle yet profound insight: two drugs that may be ineffective or even harmful for two distinct patient populations can, when the populations are combined and the drugs are randomized, guarantee an overall positive effect.

3.2.2 Mathematical Formulation

1. Patient population has p proportion of genetic makeup H, and $(1-p)$ proportion of genetic makeup T. H and T are mutually exclusive.
2. The mean response of Drug A is μ_{HA} in makeup H patients and μ_{TA} in makeup T patients.
3. The mean response of Drug B is μ_{HB} in makeup H patients and μ_{TB} in makeup T patients.
4. If we randomly select Drug A or B with probability q and $(1-q)$, respectively, what is the expected response (μ) of the population?

The expected response is the weighted average of the drug effects on the two populations, with weights determined by the corresponding sizes of each patient group:

$$\mu = pq\mu_{HA} + (1-p)q\mu_{TA} + p(1-q)\mu_{HB} + (1-p)(1-q)\mu_{TB},$$

$$\mu = \left[\left(\mu_{HA} - \mu_{TA} - \mu_{HB} + \mu_{TB}\right)q + \left(\mu_{HB} - \mu_{TB}\right)\right]p + \left(\mu_{TA} - \mu_{TB}\right)q + \mu_{TB}.$$

In practice, p is often not precisely unknown due to the invasiveness or imprecisionness of the screen tool, or no such screening tool is available or affordable. Even with a right screening tool, it is still impossible to screen the entire patient population and get precisely the patient composition. Therefore, it is desirable to choose q so that μ will be positive regardless of the unknown p. This requires that the coefficient of p for the treatment effect μ be equal to zero:

$$\left(\mu_{HA} - \mu_{TA} - \mu_{HB} + \mu_{TB}\right)q + \left(\mu_{HB} - \mu_{TB}\right) = 0,$$

Solving the equation for q, we obtain, $q = \dfrac{\mu_{TB} - \mu_{HB}}{\mu_{HA} - \mu_{TA} - \mu_{HB} + \mu_{TB}}$, where $\mu_{HA} - \mu_{TA}$ and $\mu_{TB} - \mu_{HB}$ represent the two mean differences between populations with drugs A and B, respectively, whereas $\mu_{HA} - \mu_{TA} - \mu_{HB} + \mu_{TB}$ is the sum of the mean differences.

We can rewrite,

$$q = \frac{1}{1 + \dfrac{\mu_{HA} - \mu_{TA}}{\mu_{TB} - \mu_{HB}}}, \quad \text{where} \quad 0 \le q \le 1, \tag{3.1}$$

and a positive expected treatment effect is desired:

$$\mu = \left(\mu_{TA} - \mu_{TB}\right)q + \mu_{TB} > 0. \tag{3.2}$$

From (3.1) and $q \ge 0$, we know $\dfrac{\mu_{HA} - \mu_{TA}}{\mu_{TB} - \mu_{HB}} > -1$. From (3.1) and $q \le 1$, we know

$$\frac{\mu_{HA} - \mu_{TA}}{\mu_{TB} - \mu_{HB}} \ge 0. \tag{3.3}$$

Let us refer to (3.3) as the randomizable requirement regarding the two subpopulations.

From (3.2), we obtain the positive-effect requirement (without loss of generality, assume $\mu_{TA} > \mu_{TB}$):

$$q > \frac{\mu_{TB}}{\mu_{TB} - \mu_{TA}} \quad \text{if } \mu_{TA} > \mu_{TB}; \quad q < \frac{\mu_{TB}}{\mu_{TB} - \mu_{TA}}, \quad \text{otherwise,} \tag{3.4}$$

3.2.3 Numerical Example

It is interesting that the target patient population can have different ways to divide into subpopulations. Thus, the different scenarios presented in Table 3.1 could represent different categorizations of the population, where the mean of means, $v = \dfrac{1}{4}(\mu_{HA} + \mu_{TA} + \mu_{HB} + \mu_{TB})$, represents the average treatment effect in a clinical trial when patients are equally randomized into the four groups (two treatment groups by two population groups). When μ_{HA} and μ_{TA} have different signs and the population composition (H and T) is unknown, drug A is considered ineffective or not proven effective for the overall population and thus cannot be approved for marketing. The same can be said for drug B. In this sense, many scenarios in Table 3.1 represent the cases where two ineffective drugs are made into an effective treatment with $\mu > 0$ by optimizing the randomization ratio q, assuming a sufficient sample size.

It is interesting to note that the target patient population can be divided into subpopulations in various ways. The different scenarios presented in Table 3.1 represent possible categorizations of the population. In these scenarios, the mean of means, $v = \dfrac{1}{4}(\mu_{HA} + \mu_{TA} + \mu_{HB} + \mu_{TB})$, represents the average

TABLE 3.1

Expected Treatment Effect with Optimal Randomization

Mean of means, v	μ_{HA}	μ_{TA}	μ_{HB}	μ_{TB}	Probability of getting Drug A: q	Expected effect, μ
0	−10	5	5	0	0.250	1.250
	−8.0	5	5	−2	0.350	0.450
	0.5	0	−1	0.5	0.750	0.125
	9.0	−4	−10	5	0.536	0.180
	16	−8	−24	16	0.625	1.000
	−8.0	16	16	−24	0.625	1.000
	−24	16	16	−8	0.374	1.000
0.125	1.0	0	−1	0.5	0.600	0.200
	0.5	0	−1	1	0.800	0.200
	1.25	−0.25	−1	0.5	0.500	0.125
−0.100	−10	5	5	−0.4	0.265	1.029

treatment effect in a clinical trial where patients are equally randomized into four groups (two treatment groups across two population subgroups).

When μ_{HA} and μ_{TA} have different signs and the population composition (proportions of H and T) is unknown, drug A is considered ineffective or unproven for the overall population, making it unsuitable for approval or marketing. The same applies to drug B. In this context, many scenarios in Table 3.1 illustrate cases where two individually ineffective drugs can form an effective treatment ($\mu > 0$) by optimizing the randomization ratio q, assuming a sufficiently large sample size.

For example, given $\mu_{HA} = 2$, $\mu_{TA} = -1$, $\mu_{HB} = -3$, and $\mu_{TB} = 2$; thus $v = 0$, $q = 1/(1 + \frac{3}{5}) = 0.625$. Using the optimal randomization, we expect a positive treatment effect, $\mu = 0.125$. More examples with positive treatment effects are listed in Table 3.1 for $v = 0, 0.125$, and −0.1. In all these cases, $\mu \geq v$.

It is significant that two drugs can be combined using randomization to create an effective treatment. This works because the drugs have differential effects on different populations, likely due to distinct mechanisms of action or other factors.

But why not simply match each drug to the appropriate population using diagnostic tools? The answer lies in the limitations of such tools—they may be unavailable, insufficiently accurate, too invasive, or too costly for large-scale use in general populations.

Diagnostic tools can be categorized as complementary or mandatory depending on the context of their use in medical diagnostics, the specific disease or condition being investigated, and the clinical guidelines that govern their application.

Mandatory diagnostic tools are essential for confirming a diagnosis. Without these, a definitive diagnosis cannot be made, or it may be unsafe to proceed with treatment. In contrast, complementary diagnostic tools are used in addition to mandatory tests to provide further insight into a patient's condition, but they are not strictly required for a basic diagnosis. They help in enhancing the understanding of the disease, tailoring treatment plans, or determining the prognosis.

In medical practice, the classification into mandatory or complementary often depends on specific guidelines that are informed by the latest research, expert consensus, and regulatory policies. These tools collectively aim to ensure that patients receive a diagnosis that is as accurate and detailed as necessary for effective treatment.

An additional justification for combining two ineffective drugs using randomization would be as follows: The effectiveness of this randomized combination treatment is based on its overall benefit to the population, even though it does not guarantee a positive effect for every individual. This aligns with current regulatory criteria for drug approval, which prioritize population-level efficacy over individual guarantees.

3.3 Small Data Challenges and Bayesian Optimal Randomization

An intriguing thought arises: Since the proportions of the populations (H and T) make p irrelevant to the treatment effect, it might seem unnecessary to define the patient population using inclusion and exclusion criteria. However, this idea is counterintuitive.

Upon further analysis, the missing piece becomes clear: without defining the population, we cannot determine the drug effects (μ_{TA} and μ_{TB}) and, consequently, the overall expected treatment effect:

$$\mu = (\mu_{TA} - \mu_{TB})q + \mu_{TB}.$$

Without this information, we cannot assess whether the treatment effect is positive or negative.

Even if we define the patient populations using demographics and baseline characteristics (inclusions and exclusions) from relevant clinical trials, you still do not know the true effects of the drugs. Thus, in practice, we may use the mean effects to estimate the parameters and to determine randomization:

$$\hat{q} = \frac{\underline{x}_{TB} - \underline{x}_{HB}}{\underline{x}_{HA} - \underline{x}_{TA} - \underline{x}_{HB} + \underline{x}_{TB}}.$$

The distribution of $\hat{q}$ is complex, but we can use simulations to approximate it. Practically, we may track $\hat{q}$ within the range of 0.2 and 0.8 since it is not meaningful to do it outside this range.

Under regularity conditions, for larger sample size, $\hat{q} \approx q = \dfrac{1}{1 + \dfrac{\mu_{HA} - \mu_{TA}}{\mu_{TB} - \mu_{HB}}}$.

With these random q, the mean treatment is

$$\underline{x} = (\mu_{TA} - \mu_{TB})\hat{q} + \mu_{TB}.$$

The expected treatment effect is $E(\underline{x}) = (\mu_{TA} - \mu_{TB})E(q) + \mu_{TB}$ as sample size $\to \infty$.

We do not know the expected value $E(q)$, but can use simulations to determine the sample size required to have a reasonable assurance of a positive expected treatment effect $E(\underline{x})$.

If a large sample size is not feasible, a Bayesian approach might be used. In such cases, we need to determine the Bayesian posterior distribution of randomization ratio p, incorporating subjective or objective priors.

Subjective priors incorporate specific prior knowledge or beliefs about the parameters. These priors are based on historical data (empirical priors), expert opinion (elicited priors), or other relevant information about parameter values (Informative prior).

Objective priors (non-informative priors), such as uniform, Jeffreys, and reference priors, aim to minimize the influence of prior assumptions in an analysis. Jeffreys prior is invariant under reparameterization (meaning it remains the same regardless of how the parameter is transformed) and often used because it provides a measure of objectivity, for example, beta (0.5, 0.5) for a proportion. Reference priors are constructed to maximize the expected information gain from the data and can be used to produce more objective Bayesian inferences.

Objective priors are based on mathematical properties (e.g., invariance and uniformity), while subjective priors are based on specific prior information or beliefs.

The beta distribution is a continuous probability distribution defined on the interval [0,1]. It is parameterized by two positive shape parameters, α and β, to determine the shape of the distribution as follows:

$$f(x : \alpha, \beta) = \frac{x^{\alpha-1}(1-x)^{\beta-1}}{B(\alpha, \beta)}, \text{ if } 0 \leq x \leq 1, \text{ otherwise, } 0.$$

Here $B(\alpha, \beta) = \dfrac{\Gamma(\alpha)\Gamma(\beta)}{\Gamma(\alpha+\beta)}$.

For a positive integer n, $\Gamma(n) = (n-1)!$

The mean is $\mu = \dfrac{\alpha}{\alpha+\beta}$ and variance $\sigma^2 = \dfrac{\alpha\beta}{(\alpha+\beta)^2(\alpha+\beta+1)}$.

Uniform distribution on [0, 1] is a special case when $\alpha = \beta = 1$.

3.4 Response-Adaptive Randomization for Clinical Trials

3.4.1 Dynamic Patient Allocation Inspired by Clot Machine Game

Imagine you are playing two slot machines and suspect one might have a slightly higher chance of winning than the other. You could apply a strategy known as play-the-winner: If you win on Machine A (or B), continue playing that same machine; if you lose, switch to the other machine. Repeat this process until you decide to stop. This strategy increases your chances of winning by favoring the machine with cumulative successes.

We now draw an analogy between slot machines and medical treatment options with unknown response rates, where winning corresponds to successful treatment outcomes, forming the basis for response-adaptive clinical trial designs. However, due to the blinding requirement in clinical trials, the deterministic switching rule is modified to a probabilistic approach. One common model to adjust treatment-switching probability is the randomized play-the-winner (RPW) design. This model further biases allocation toward better-performing treatments by directly incorporating success/failure into ball adjustments (Figure 3.2).

3.4.2 RPW Rules

1. Initialization: Start with *k* balls for each treatment in the urn.
2. Patient allocation: Randomly select a ball to determine the treatment arm.

Play-the-Winner Rule

1. **Initialization**: Start with *k* balls for each treatment in the urn.
2. **Patient Allocation**: Randomly select a ball to determine the treatment arm.
3. **Outcome Update**:
   - If the outcome is a **success**, add *m* balls of the chosen treatment's color.
   - If the outcome is a **failure**, add *m* balls for the alternative treatment.
4. **Dynamic Adaptation**: Over time, treatments with higher success rates dominate the urn, leading to preferential assignment of patients to those arms.

FIGURE 3.2
Play-the-winner rules for clinical trial design.

3. Outcome update:
   - If the outcome is a success, add m balls of the chosen treatment's color.
   - If the outcome is a failure, add m balls for the alternative treatment.
4. Dynamic adaptation: Over time, treatments with higher success rates dominate the urn, leading to preferential assignment of patients to those arms.

This setup demonstrates how the RPW rules adaptively allocate resources to the better-performing treatment in real time, optimizing patient outcomes within the trial framework. The idea of response-adaptive trials can be generalized to a multiple arms trial, in which the successful probability of each arm is unknown (Chang, 2007, 2014).

3.4.3 Features of Response-Adaptive Clinical Trials

In the RPW model, sample size ratio and response rates in the two groups are asymptotically related: the sample size ratio between two groups n_1/n_2 asymptotically approaches the ratio of their response rates p_1/p_2 under consistent response patterns. The sensitivity and speed of this convergence depend on the initial allocation (k balls) and the updating parameter (m balls).

In other words, the initial phase involves exploration (allocating patients across treatments to collect response data), while later phases focus on exploitation (allocating more patients to the treatment that shows better results). This balance is crucial: too much exploration (a small value of m/k) may expose more patients to less effective treatments, and too much exploitation (a large value of m/k) can lead to biased or unreliable results with limited sample size.

In the slot machine analogy, the goal is to maximize cumulative rewards over time. The player wants to maximize their winnings within a set number of plays. Similarly, in response-adaptive trials, the goal is to maximize patient benefit, often by ensuring that as many patients as possible receive the more effective treatment. The adaptive mechanism works to identify and allocate more patients to the better-performing treatment arm, maximizing cumulative benefits for trial participants in terms of the specified clinical endpoint such as relative risk, proportion, or odds ratio. Thus, it is an optimal design.

Response-adaptive trials are increasingly employed in various areas of clinical research. In oncology trials, they are particularly valuable in early-phase studies where the rapid assessment of efficacy and safety is critical. In rare disease research, where patient numbers are often limited, these designs help maximize the information gained from each individual patient. Similarly, in emergency situations such as infectious disease outbreaks, response-adaptive trials play a vital role in quickly identifying effective treatments.

For a more detailed exploration of response-adaptive trial designs, readers are encouraged to consult foundational and advanced works, including those by Rosenberger et al. (2012), Hu and Rosenberger (2006), Zelen (1969), Wei (1978), Ivanova (2003), and Chang (2014).

3.5 Summary: Harnessing Randomization for Clinical Trial Designs

This chapter explores the concept of randomization in decision-making through critical thinking and creative analogies, illustrating its application in clinical trials. By leveraging the fair-game paradox as an analogy, it demonstrates how strategic randomization can turn two individually ineffective treatments into an effective combined therapy without physical or chemical modifications. This chapter emphasizes the role of randomization in optimizing outcomes, particularly in challenging scenarios such as small sample sizes or heterogeneous populations.

Key Takeways:

1. Fair-game paradox as a framework: The paradox illustrates how altering the arrangement of decisions (e.g., coin flips or patient allocations) can lead to optimized outcomes. The analogy serves as a foundation for understanding how randomization strategies enhance decision-making in clinical contexts.

2. RPW Model: Borrowing from the "clot machine" analogy, the RPW model dynamically adjusts treatment allocations in response-adaptive clinical trials, favoring treatments with better observed success rates. This approach balances ethical concerns (minimizing patient exposure to less effective treatments) with statistical rigor.

3. Implications for clinical trials: The chapter highlights the ability of randomization to overcome diagnostic limitations, such as imprecise screening tools, by relying on probabilistic frameworks to optimize overall treatment effects.

This chapter highlights the intersection of logic, creativity, and practical application in advancing clinical trial design and decision-making strategies.

Exercise

1. Explain the fair-game paradox and how the arrangement of coins by Player B influences the expected outcomes. Use the equation $G = -10ab + 5a(1-b) + 5b(1-a)$ to illustrate your explanation.

2. Describe how randomization strategies inspired by the fair-game paradox can combine two ineffective drugs into an effective treatment. Why is this approach significant for clinical trial design?

3. What critical thinking ideas and elements of creative analogies stand out to you in this chapter?

4. Explain the RPW model. Use computer simulations to analyze the effects of urn parameters—including the initial numbers of black and white balls (k) and the number of balls added after each success (m)—on the patient distribution between the two treatments and the variability in success rates for each treatment.

5. Using the formula $q = \dfrac{\mu_{TB} - \mu_{HB}}{\mu_{HA} - \mu_{TA} - \mu_{HB} + \mu_{TB}}$, calculate the randomization ratio q for a scenario where $\mu_{HA} = 2$, $\mu_{HB} = -3$, $\mu_{TA} = -1$, and $\mu_{TB} = 2$. Discuss the implications of your result.

6. Develop an analogy to explain response-adaptive randomization to a lay audience. Use a real-world example, such as choosing a playlist based on listener preferences.

7. Propose an analogy that explains how randomization can combine two ineffective treatments into an effective one. Relate this analogy to a non-medical context, such as teamwork in sports.

8. Debate whether it is ethical to prioritize population-level benefits over individual outcomes in clinical trials using randomization strategies. What safeguards should be implemented?

9. Compare Bayesian and frequentist methods for optimizing randomization strategies in clinical trials with small sample sizes. Which approach is more practical, and why?

10. Explore the dynamics of the play-the-winner model with more than two treatment options. How would the randomization strategy change, and what factors need to be considered?

11. Study the effect of parameters k and m in the play-the-winner model when chaos happens, i.e, when there is a good chance to pick the loser or inferior one? A probability of picking a loser as a function k and m or a simulated plot is preferred.

12. Why is a drunk randomly choosing a key from a set of keys in his hand more likely to succeed on his first attempt? Does this imply

that most people are also more likely to succeed on their first attempt at starting a company?

13. Extend the method described in Section 3.2 to address the following scenarios: (1) trials with two co-primary endpoints, (2) trials with more than two treatments and subpopulations, and (3) applicability to binary endpoints.

4

An Anti-Polarization Political System Inspired by the Play-the-Winner Strategy

At the heart of polarization lies a paradox: the louder we assert our truths, the harder it becomes to hear anyone else's.

Polarization feeds on the illusion of absolutes, blinding us to the nuanced truths that exist in the gray.

True polarization begins not with disagreement, but with the unwillingness to understand.

4.1 Democratic Systems

4.1.1 Elements of Democratic System

A democratic system is a form of government in which power is vested in the people, allowing them to have a direct or indirect say in decisions that affect their lives. In a democracy, leaders are chosen through free and fair elections, and policies are influenced by the preferences of the citizenry, either directly through referendums or indirectly through representatives. The fundamental principles of a democratic system include participation, equality, accountability, and the rule of law. Here is an outline of the essential characteristics of a democratic system:

1. Political participation: In a democracy, citizens have the right to participate in political processes, especially by voting in elections to choose their representatives. This also includes freedoms of speech, assembly, and the press, allowing people to advocate for issues, protest, and hold leaders accountable.

2. Free and fair elections: A hallmark of democracy is regular, free, and fair elections, where citizens can vote for their leaders without coercion or manipulation. Elections must be open to all eligible citizens, and all votes should have equal weight, ensuring the principle of "one person, one vote."

DOI: 10.1201/9781003630081-4

3. Rule of law: In a democratic system, the rule of law means that laws apply equally to all citizens, including leaders and government officials. No one is above the law, and the legal system is expected to be fair, transparent, and impartial.

4. Protection of rights and liberties: A democratic system guarantees certain fundamental rights and freedoms, such as freedom of speech, religion, press, assembly, and the right to a fair trial. These rights are often enshrined in a constitution or legal framework that limits government power and protects individual freedoms.

5. Separation of powers and checks and balances: To prevent abuses of power, democracies typically divide government into branches, such as the executive, legislative, and judicial branches. Checks and balances ensure that no single branch or individual has unchecked power, creating accountability among the branches.

6. Accountability and transparency: Democratic leaders are accountable to the people, meaning they must answer to citizens for their actions and decisions. Transparency in government operations, such as public access to information and open policy discussions, is key to fostering trust and accountability.

7. Pluralism and tolerance: Democracies encourage a diversity of opinions, allowing different groups and individuals to express their beliefs and participate in public life. Political pluralism means that multiple political parties and ideologies can coexist, compete, and contribute to the governing process.

8. Responsiveness to public needs: In a democratic system, government officials are expected to respond to the needs and interests of the people. Policy decisions should reflect the will of the majority while also protecting minority rights, ensuring that governance serves the interests of the populace as a whole.

4.1.2 Types of Democratic Systems

Democracy can take several forms depending on how citizens participate and how power is structured:

- Direct democracy: Citizens directly vote on policy decisions rather than through representatives. This system is most feasible in smaller or local contexts, such as town hall meetings or referendums.

- Representative democracy: Citizens elect representatives to make decisions on their behalf. This is the most common form of democracy globally, as it is more practical for larger populations.

- Parliamentary democracy: The executive branch is derived from the legislative branch (parliament), and the head of government is often a prime minister chosen by parliament members.
- Presidential democracy: The executive branch is separate from the legislature, and the president is elected independently by the people, serving as both head of state and government.

The benefits of democratic systems include promoting political and civil freedoms, creating a government responsive to the will of the people, providing mechanisms for peaceful transitions of power and dispute resolution, and fostering accountability and transparency. However, democratic systems also face challenges, such as susceptibility to populism, where leaders appeal to emotions rather than reason, the potential for polarization and divisive politics in highly pluralistic societies, and the reliance on a well-informed electorate, which can be hindered by misinformation or limited access to education.

In summary, a democratic system emphasizes the people's right to influence governance, either directly or indirectly, ensuring leaders are accountable, rights are protected, and power is balanced. It is not only a framework of institutions but also a commitment to the principles of justice, equality, and freedom that empower society to shape its collective future.

4.2 Trends and Impacts of Political Polarization

Political polarization has undergone significant changes globally, influenced by various social, economic, technological, and cultural factors. Here is an overview of key trends and changes in political polarization over recent decades:

1. **Intensification of Ideological Divides**
   - Partisan alignment: Political parties have increasingly aligned along stark ideological lines. For instance, in countries like the U.S., Democrats and Republicans have grown further apart on issues like healthcare, climate change, and immigration.
   - Cultural polarization: Social and cultural issues—such as gender rights, racial equity, and religion—have become central points of contention, deepening divides.
2. **The Role of Technology and Media**
   - Social media's influence: Platforms like Twitter, Facebook, YouTube, and Cable TVs amplify echo chambers, where people are exposed to viewpoints similar to their own, reinforcing existing biases.

- Misinformation: Polarization is exacerbated by the spread of fake news, conspiracy theories, and targeted disinformation campaigns.
- Fragmentation of traditional media: Media outlets catering to specific political ideologies have proliferated, further fragmenting audiences.

3. **Economic and Class Factors**
   - Inequality: Rising economic inequality has deepened divides, particularly in advanced economies. Economic grievances often align with political divides, as seen in populist movements.
   - Globalization's impact: Perceptions of lost jobs and cultural identity due to globalization have fueled polarization, with populist leaders often capitalizing on these sentiments.

4. **Cross-National Comparisons**
   - Western democracies: Countries like the U.S. and the UK have experienced significant polarization due to cultural wars and economic inequality.
   - Emerging economies: Nations such as India and Brazil face polarization along ethnic, religious, and socio-economic lines.
   - Authoritarian contexts: Polarization in some countries often involves suppression of dissent rather than open democratic divides.

5. **Counter-Trends and Responses**
   - Coalition building: Efforts to form bipartisan coalitions, as seen in the EU on climate policies, highlight attempts to bridge divides.
   - Civic engagement: Grassroots movements aimed at fostering dialogue (e.g., deliberative democracy initiatives) show promise in mitigating polarization.
   - Media literacy campaigns: Educating the public about misinformation and bias in media can reduce susceptibility to divisive narratives.

4.2.1 Key Implications of Increased Polarization

- Policy gridlock: Deep divides make consensus-building difficult, often leading to legislative paralysis.
- Social cohesion: Polarization erodes social trust, increasing tensions and, in extreme cases, contributing to civil unrest.
- Global impact: Polarization within major powers can affect international cooperation on critical issues like climate change, trade, and security.

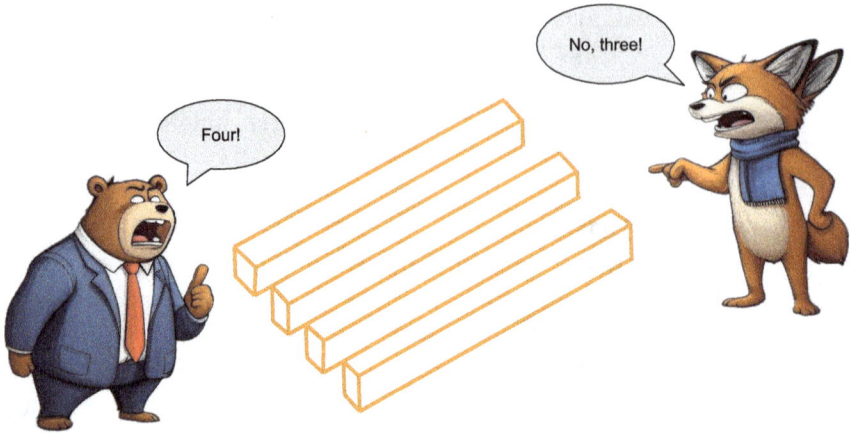

FIGURE 4.1
Polarization—intensified by the reluctance to acknowledge other perspectives.

Understanding and addressing political polarization demands a multifaceted approach, encompassing education, media reform, economic inclusion, and open dialogue (Figure 4.1). However, beyond these conventional strategies, exploring innovative changes in voting systems and political governance—drawing inspiration from the response-adaptive randomization (RPW) in adaptive clinical trials—could significantly reduce polarization. This concept is the focus of the next two sections.

4.3 Voting Strategy to Reduce Political Polarization

Faced with the increasing polarization of U.S. society, changes in election laws could help promote moderation and bridge divides. One proposed strategy involves encouraging voters to support candidates from different parties during the primary and general elections. For example, if you plan to vote for a Democrat in the general election, you would be encouraged to vote for a Republican in the primary. This approach incentivizes voters to select a moderate Republican or one whose views align more closely with Democrats, increasing the likelihood that both major parties nominate centrist candidates. As a result, the general election would feature a choice between two moderate, broadly appealing candidates.

This strategy emphasizes "encouragement" rather than a mandatory requirement. By allowing voters to maintain autonomy while guiding them toward choices that favor moderation, this method respects individual preferences while addressing societal polarization. Encouraging cross-party voting in this way could help foster collaboration and reduce the ideological extremes that often dominate the political landscape.

4.3.1 How Encouraging Cross-Party Voting Could Work

1. Incentivizing cross-party voting: Instead of requiring voters to split their choices, the system could offer incentives or simply raise public awareness about the benefits of voting for a moderate candidate from the opposite party in the primary. This could be achieved through voter education campaigns, policy discussions, or even modest, voluntary rewards like a slight tax credit or recognition for cross-party voting participation.

2. Promoting moderate candidates: By encouraging voters to select a candidate from the opposite party in the primary, they are more likely to support a candidate closer to the center. This can help bring moderate Republicans and Democrats to the forefront, making it more likely that the final election features candidates with broader appeal and centrist policies.

3. Reducing negative partisanship: Encouragement rather than compulsion preserves voter choice, making it more likely that people will try it voluntarily. This approach also reduces the adversarial mindset where voters see the other party as "the enemy" and instead fosters a spirit of bipartisan consideration.

4.3.2 Advantages of Encouragement over Requirement

1. Respects voter autonomy: By not mandating cross-party voting, this strategy avoids infringing on the freedom of choice that is foundational to democracy. Voters who feel strongly aligned with one party are not forced to support another party, but they are encouraged to consider it.

2. Lower resistance and higher buy-in: Voter encouragement rather than compulsion is likely to face less public resistance. Mandatory rules tend to be controversial, especially in the context of voting rights. By promoting cross-party voting as an option rather than a requirement, this strategy is more likely to receive bipartisan support.

3. Supports an informed electorate: Encouraging voters to consider candidates from both parties could lead to a more informed electorate. Exposure to a broader range of candidates and ideas might help voters make decisions based on policy rather than party loyalty alone.

4. Flexibility to experiment: Encouragement can be implemented as a pilot program or targeted campaign, allowing time to measure its effectiveness. If it shows promising results in terms of reducing polarization, it can be expanded or refined without immediately reshaping election laws.

4.3.3 Possible Mechanisms for Encouragement

1. Public awareness campaigns: Government or nonpartisan organizations could create campaigns emphasizing the benefits of considering candidates from both parties, particularly in primaries. Messaging could highlight how this practice might reduce polarization and lead to more collaborative governance.

2. Educational programs: Voter education materials could include information on each candidate's policy stances, especially focusing on cross-party appeal. This might lead more voters to consider moderate candidates who align with their values, regardless of party.

3. Positive recognition: While it may not be feasible to offer direct incentives for cross-party voting, positive recognition (such as labeling certain candidates as "bipartisan choices") could influence voters. This would provide a social cue that voting across party lines is beneficial and valued.

4. Primary structure reforms: Adjusting the primary system to be more open or adopting "top-two" primary formats, where all candidates compete in a single primary with the top two advancing, could complement encouragement efforts by giving voters from both parties more say in which candidates move forward.

4.3.4 Remaining Challenges

While encouragement could be a powerful approach, it may still face a few challenges:

1. Overcoming partisan loyalty: Strong partisan loyalties mean that some voters may be resistant to voting for any candidate from the opposing party, even in the primary. Persuading these voters to consider moderation may require sustained and nuanced educational efforts.

2. Risk of ineffectiveness: Encouragement without strong incentives might not produce a significant shift in voter behavior. Therefore, careful monitoring and iterative improvements would be necessary to measure and enhance the impact of this approach.

3. Potential strategic voting: Encouraging cross-party voting could still lead to strategic voting, where people vote for the least viable candidate in the opposing primary to weaken that party's options. This risk could be mitigated by focusing educational efforts on the importance of constructive participation in both parties' primaries.

4.3.5 Final Thoughts

Using encouragement instead of compulsion is a more feasible and democratic way to foster cross-party engagement and reduce polarization. While not a guaranteed solution, it could gradually shift voter behavior toward favoring moderate candidates. If done thoughtfully, this approach has the potential to make primaries less partisan and increase the likelihood of electing centrist candidates with broader appeal. This aligns well with democratic values of choice, respect for diversity of opinion, and fostering a more informed electorate.

Another potential voting strategy to reduce polarization is the "one person, two votes" system. In this approach, each individual casts two votes: one positive vote for a preferred candidate and one negative vote against a less favored candidate. A candidate's overall qualification is determined by the net difference between their positive and negative votes.

4.4 Anti-Polarization Democratic System

4.4.1 Overview of the Proposed System

Current winner-takes-all system: In many presidential systems, the candidate who receives the majority of votes wins the entire executive power for a fixed term. Thus, only one party governs, even if the majority is slim, leaving minority parties without executive power.

To reduce the polarization, we draw an analogy between a play-the-winner game and the political system, proposing a new political election framework centered on proportional governing time. Under this system, voters cast their ballots for their preferred candidates or parties, and the total governing period is allocated proportionally among parties based on their share of the vote. The votes can be individual-based or electoral-based, but an individual-based system is preferable, as proportional governing time better reflects equality among individuals and states.

4.4.2 Example

- Total governing period: 10 years.
- Election results:
  - Candidate A (Party A): 50% of votes ⇒ 5 years of governance.
  - Candidate B (Party B): 30% of votes ⇒ 3 years of governance.
  - Candidate C (Party C): 20% of votes ⇒ 2 years of governance.

- Minimum governing time: If a party's allocated time is less than a set minimum (e.g., 3 years), the time can be carried over to the next term.
  - Party C's situation: Allocated 2 years, carries over to next term.

4.4.3 Adjustments and Variations

- Election frequency: Elections could occur every 10 years or another interval, adjusting governing times accordingly.
- Minimum governing threshold: Setting a minimum period ensures parties have sufficient time to implement policies.
- Carryover mechanism: Unused governing time can be added to the party's allocation in the next cycle.

4.4.4 Potential Advantages

1. Proportional Representation
   - Inclusivity: More accurately reflects the electorate's preferences in governance.
   - Minority representation: Smaller parties gain executive power, allowing them to contribute to policy-making.
2. Reduction in Polarization
   - Collaboration incentives: Parties may be encouraged to adopt more moderate positions to appeal to a broader electorate.
   - Decreased zero-sum dynamics: Since losing parties still gain governing time, the high-stakes competition that fuels polarization may diminish.
3. Policy Continuity and Diversity
   - Variety of policies: Different parties bring varied policy perspectives, enriching governance.
   - Long-term planning: Extended governing periods allow for the implementation of long-term policies.
4. Less Frequent Election
   - Less frequent election can reduce "election fatigue syndrome."

4.4.5 Legislative Aspects

Legislative oversight can strengthen the role of the legislature in providing continuity. However, polarization between the two major political parties often manifests in disagreements over the fiscal year budget, which can lead to government shutdowns. These shutdowns disrupt essential services, undermine public trust, and damage the economy. To address this issue, a "third-party rule" could be implemented to maintain government operations and reduce legislative gridlock.

Under the third-party rule, items agreed upon by both parties would remain unchanged. For contentious items, a third popular party would make the final decisions, operating within the boundaries of the two parties' proposals to ensure a balanced resolution.

Additionally, strict deadlines for budget approval could be enforced with penalties for lawmakers. For example, withholding their salaries during a shutdown would serve as a powerful incentive to encourage timely decision-making, promote accountability, and minimize legislative delays. While the proposed system is novel, existing political systems, such as Switzerland's Federal Council, offer valuable insights. The Federal Council operates as a collective executive body, where a multi-member council represents major political parties. The presidency rotates annually among council members, ensuring shared leadership responsibilities. This structure fosters consensus-building and helps reduce polarization within the government.

4.4.6 Broad Clemency Power

The U.S. Constitution grants the president nearly unlimited authority to pardon individuals for federal offenses, with the sole exception of cases involving impeachment. This expansive power includes the ability to issue preemptive pardons for crimes committed up to a specified date. However, this authority operates without meaningful oversight. As political polarization deepens, there is a growing concern that a president might exploit this power to protect allies, secure political leverage, or make controversial promises, such as offering preemptive pardons as part of a campaign strategy. In extreme scenarios, this power could theoretically be abused to shield individuals involved in politically motivated acts, such as violence or even assassination.

To address the potential for such misuse, several strategies could help establish safeguards and accountability. Transparency could be enforced through a mandate requiring public disclosure and detailed justifications for all pardons. This would ensure that the rationale behind each pardon is subject to scrutiny. Prohibiting preemptive pardons for crimes that have not yet been formally charged or prosecuted is another potential solution, as is restricting the issuance of pardons during active election campaigns to prevent their use as a political tool. Additionally, a system could be implemented allowing Congress to override a presidential pardon if more than two-thirds of senators formally object. These measures could provide a necessary check on this significant power while maintaining the president's constitutional authority.

4.5 Summary: Proportional Governance to Counter Political Polarization

This chapter explores a novel approach to reducing political polarization inspired by the RPW rule, originally developed for adaptive clinical trials. It proposes a proportional governance system, where governing time is allocated based on the proportion of votes each party receives, rather than the winner-takes-all approach commonly used in presidential systems. By ensuring that multiple parties share executive power, it seeks to reflect the electorate's diverse preferences, promoting inclusivity and fairness while reducing the high-stakes competition inherent in winner-takes-all systems. The Third-Party Rule can potentially reduce the likelihood of government shutdowns. Imposing numerical or sentence-year limits on presidential pardons could reduce the risk of abuse and improve public perception of the clemency process.

Key Takeaways:

1. Identifying flaws in the current system: This chapter critically examines the winner-takes-all model, highlighting its tendency to marginalize minority parties and exacerbate polarization.

2. Adapting the RPW rule: Borrowing from the adaptive allocation strategies in clinical trials, the chapter reimagines political governance as a proportional system to optimize outcomes for a diverse electorate.

3. Practical adjustments: Mechanisms like minimum governing thresholds and carryover of unused governing time offer flexibility to maintain stability.

4. Evaluating feasibility: Challenges such as implementation complexity, voter acceptance, and strategic adjustments by parties are acknowledged, demonstrating a balanced, analytical perspective.

Exercise

1. Comparative analysis: Compare the "play-the-winner political system" proposed in this chapter to the existing "winner-takes-all" system in terms of inclusivity and polarization. Identify at least three strengths and three weaknesses of each system.

2. Proportional governing time: If an election results in 45%, 35%, and 20% of the votes for three parties, calculate the proportional governing times over a 10-year period. Discuss how the proposed system could influence voter behavior and party strategies.

3. What critical thinking ideas and elements of creative analogies stand out to you in this chapter?

4. Reducing polarization: The chapter suggests encouraging cross-party voting. Develop a detailed proposal for an educational campaign aimed at promoting this idea. Include potential challenges and ways to address them.

5. Global context: Research a country that uses proportional representation (e.g., Switzerland or New Zealand). Compare their system with the play-the-winner framework. What lessons can be learned from these countries to refine the proposed system?

6. Analogy exercise: This chapter uses an analogy between the play-the-winner model and clinical trial randomization. Create your own analogy to explain the proportional governing system in a way that is relatable to students from diverse disciplines.

7. Game theory: Design a simple game or simulation to demonstrate how proportional governing time might work in practice. Include rules, scenarios, and outcomes that show its potential advantages and challenges.

8. Debate exercise: Divide into two groups. One group argues for the adoption of the play-the-winner system in a major democracy like the U.S., whereas the other group defends the existing system. Use real-world examples to support your arguments.

9. Impact on minority representation: Discuss how the proposed system might impact minority groups or smaller political parties. Would it enhance their influence, or could it lead to unintended consequences?

10. Third-party rule: This chapter suggests a "third-party rule" for resolving legislative gridlocks. Propose a detailed implementation plan for this rule in the U.S. Congress, considering its potential benefits and drawbacks.

11. Future adaptations: Imagine that the proportional governing system is implemented in a country. Describe how it could evolve over the next 50 years. What adjustments might be necessary to keep it effective and relevant?

5

Clinical Trial Design and Development: Insights from Game Theory and Stochastic Processes

Cooperative game theory analogies illuminate strategies to optimize clinical trial outcomes.

5.1 The Prisoner Paradox

Two prisoners are held in separate rooms and cannot communicate. Both understand the rules of the game: there is no loyalty between them, and there will be no opportunity for retribution or reward outside the game. The setup is as follows:

Each prisoner can choose to cooperate (do not confess) or defect (confess). Regardless of the other's decision, each prisoner gets a better outcome by defecting. This reasoning is based on analyzing the best responses for both players (Figure 5.1):

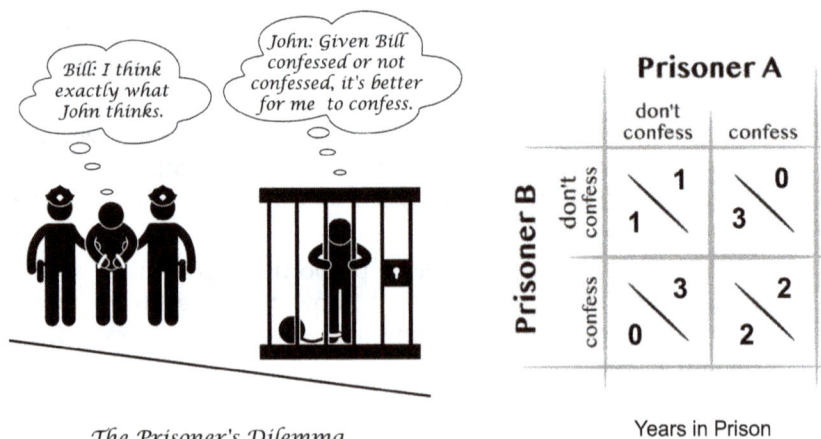

Bill: I think exactly what John thinks.

John: Given Bill confessed or not confessed, it's better for me to confess.

The Prisoner's Dilemma

Prisoner A

		don't confess	confess
Prisoner B	don't confess	1 / 1	0 / 3
	confess	3 / 0	2 / 2

Years in Prison

FIGURE 5.1
The prisoner's dilemma: a game theory classic.

DOI: 10.1201/9781003630081-5

- If Prisoner B cooperates, then Prisoner A should defect, since being set free is better than serving one year in prison.
- If Prisoner B defects, then Prisoner A should also defect, as serving two years is preferable to serving three.

Thus, defection is the optimal strategy for Prisoner A, regardless of Prisoner B's choice. The same reasoning applies to Prisoner B, making defection the dominant strategy for both players. As a result, both prisoners will choose to defect, leading to a mutual defection outcome (each serves 2 years in prison), which is the game's only Nash equilibrium—where neither player can unilaterally improve their situation by changing their choice. However, this is not the better solution for either of them: If they cooperate do not confess, each will serve 1 year instead of 2 years in prisons.

5.2 Key Concepts in Game Theory

Nash equilibrium represents a situation in which no player can improve their payoff by unilaterally changing their strategy, provided other players stick to their strategies. For example, in a pricing competition, if two firms choose optimal prices where neither can increase profit by deviating, they are in Nash equilibrium.

A dominating strategy is one that always provides at least as good a payoff as any other strategy, regardless of the opponent's actions. It consistently outperforms alternatives. When all players adopt their dominant strategies, the result is a dominant strategy equilibrium, a stable outcome. For instance, in the prisoner's dilemma, "defecting" is a dominant strategy for both players, leading to an equilibrium where both defect.

In a zero-sum game, one player's gain equals the other player's loss, resulting in a total payoff of zero. Classic examples include chess and poker, where one player's victory is the other's defeat.

A mixed strategy involves probabilistically choosing among available options rather than committing to a single strategy. In rock-paper-scissors, for example, playing each option with equal probability (one-third) constitutes a mixed strategy.

Non-cooperative solutions arise when players act independently to maximize their own payoffs without collaborating. Nash equilibrium is a typical non-cooperative solution. In contrast, cooperative solutions occur when players coordinate or form coalitions to improve collective outcomes, such as bargaining in a trade agreement to maximize joint benefits.

A Pareto optimum is a resource allocation where it is impossible to make one individual better off without making another worse off. In the prisoner's dilemma, the mutual defection outcome is not Pareto optimum, but Pareto-inefficient, as moving to mutual cooperation would be a Pareto improvement that benefits both prisoners without harming either.

The maximin criterion, proposed by von Neumann and Morgenstern (1944), applies to two-person zero-sum games, where each player chooses a strategy that maximizes their minimum payoff. The resulting pair of strategies and payoffs defines the solution to the game. Similarly, Nash's theorem (1951) asserts that every finite game—whether pure or mixed, with a finite number of players and strategies—has at least one Nash equilibrium.

However, this Nash equilibrium is not Pareto-efficient, because while mutual defection is the rational choice from a self-interested perspective, the ideal outcome—mutual cooperation —would lead to a better result for both players. This illustrates the conflict between individual rationality and collective optimality.

In a two-player game, the players effectively have separate objective functions: Player A aims to optimize $f(x, y)$ with respect to their own choice of x, and Player B aims to optimize $g(x, y)$ with respect to y. Here, x and y are independent decision variables for the two players. This differs from a common optimization problem, where a single objective function, $H(f(x, y), g(x, y))$, would be optimized with respect to both x and y, where y might be a function of x.

5.3 Clinical Trial Designs Inspired by the Prisoner Paradox

A clinical trial is a research study conducted with human participants to evaluate the safety, efficacy, and optimal use of medical interventions such as drugs, medical devices, treatments, or diagnostic procedures. It follows a structured protocol to assess outcomes, comparing the intervention under investigation with a placebo, existing treatments, or different doses. The goal is to generate evidence to support regulatory approval, guide clinical practices, or improve patient care. Clinical trials are typically conducted in phases to gradually assess safety and effectiveness in different populations and settings. Clinical trial design is a key element in the entire process.

In a clinical trial, we must consider at least two critical endpoints: safety and (primary) efficacy. Suppose retrospective safety studies indicate that the test drug is safer in Population B than Population A, regardless of dose level (low dose L or high dose H), and that Dose H is more efficacious than Dose L, irrespective of the population. Based on this information, should we automatically choose Population B and Dose H to conduct the clinical trial?

To conceptualize this decision-making process, we can draw an analogy to the Prisoner's Paradox and assign roles to the "players" in this game:

- Player 1 (efficacy guard): Chooses between Population A or Population B based on efficacy considerations.
- Player 2 (safety guard): Chooses between Dose L or Dose H based on safety considerations.
- Player 1, after conducting a retrospective safety study, reasons that whether Dose L or H is used, population B is safer and has promising efficacy signals; thus B is the choice.
- Player 2, after conducting a retrospective efficacy study, reasons that whether in population A or B, Dose H is more efficacious; thus H is the choice.

Without considering other facts such as the costs of L and H, and sizes of A and B, should the optimal solution be (Dose H, Population B) in terms of safety and efficacy? In any case, we would expect (H, B) will give us an optimal solution at least in terms of one of the criteria: safety or efficacy. However, to everyone's surprise, it is not true, in the current case, the combination (L, A) is better than (H, B) in terms of efficacy, safety, or both, as shown in Figure 5.2. Therefore, launching a large clinical trial with low dose on population A will be expected to improve both safety and efficacy scores from 1 to 2 compared to the original choice of high dose on population B.

Of course, in practice, there are more aspects to consider such as the need to establish a dose–response relationship.

The endpoints in the example do not have to be safety and efficacy endpoints; there can be other multiple endpoints such as quality of life, cost, convenience, and invasiveness.

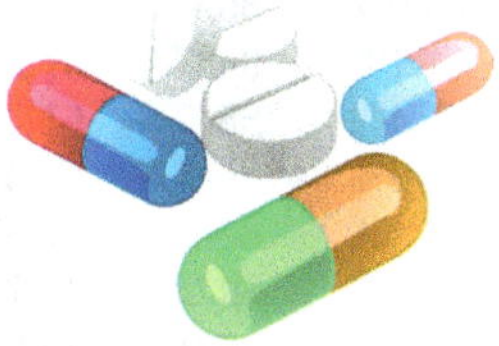

		Player 1: Safety	
		A	B
Player 2: Efficacy	L	2 \ 2	0 \ 3
	H	3 \ 0	1 \ 1

A Higher Value Indicates Safer or More Efficacious.

FIGURE 5.2
Drug safety–efficacy game payoff table.

5.4 Optimization of the Clinical Development Program

Drug development typically progresses through sequential phases, from Phase 1 to Phase 3. A successful trial advances the drug to the next phase. Upon completing Phase 3, the drug is submitted to the FDA in the United States as part of a New Drug Application (NDA) or to other international regulatory agencies for marketing approval. Phase transition probabilities (success rates) by disease area are summarized in Figure 5.3 (Chang, 2019, p. 60).

These probabilities quantify the success likelihood across the clinical development program (CDP). The process can be modeled using a stochastic process, a sequence of random variables $X_1, X_2, ..., X_n$ with associated probabilities of transitioning between states. A Markov chain, a simple stochastic process, assumes that the probability of moving to the next state depends only on the current state.

Figure 5.4 illustrates a Markov chain model for clinical success probabilities, where P_{ij} is the transition probability from state i to j. Using this model, the probability of progressing from Phase I to approval is the product of three transition probabilities

$$0.632 \times 0.307 \times 0.496 = 0.096.$$

	Test Drug	Phase 1 to Ph 2	Phase 2 to Ph 3	Phase 3 to NDA	NDA to Appr	Phase 3 to Appr
Hematology	86	73.3%	56.6%	75.0%	84.0%	63.0%
Infectious disease	347	69.5%	42.7%	72.7%	88.7%	64.5%
Ophthalmology	66	84.8%	44.6%	58.3%	77.5%	45.2%
Other	96	66.7%	39.7%	69.6%	88.4%	61.5%
Metabolic	95	61.1%	45.2%	71.4%	77.8%	55.5%
Gastroenterology	41	75.6%	35.7%	60.6%	92.3%	55.9%
Allergy	37	67.6%	32.5%	71.4%	93.8%	67.0%
Endocrine	299	58.9%	40.1%	65.0%	86.0%	55.9%
Respiratory	150	65.3%	29.1%	71.1%	94.6%	67.3%
Urology	21	57.1%	32.7%	71.4%	85.7%	61.2%
Autoimmune	297	65.7%	31.7%	62.2%	86.0%	53.5%
Neurology	462	59.1%	29.7%	57.4%	83.2%	47.8%
Cardiovascular	209	58.9%	24.1%	55.5%	84.2%	46.7%
Psychiatry	154	53.9%	23.7%	55.7%	87.9%	49.0%
Oncology	1222	62.8%	24.6%	40.1%	82.4%	33.0%
All Indications	3582	63.2%	30.7%	58.1%	85.3%	49.6%

FIGURE 5.3
Phase transition probabilities of clinical trials.

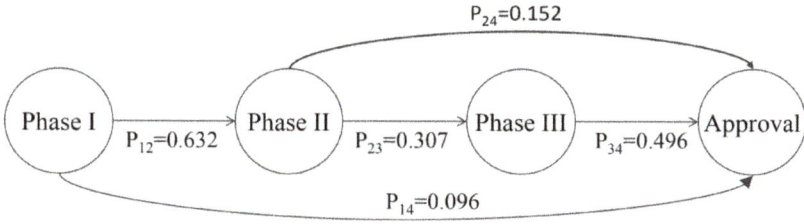

FIGURE 5.4
Markov chain for CDP.

Similarly:

- Phase II to approval: $0.307 \times 0.496 = 0.152$
- Phase III to approval: 0.496

In general, the probability of transitioning from state i to $i+k$ is:

$$P_{i,i+k} = P_{i,i+1}P_{i+1,i+2}\ldots P_{i+k-1,i+k}.$$

This raises the question: should decision rules prioritize increasing early-phase transition probabilities? Early termination of ineffective or toxic candidates is desirable to minimize resource waste. However, there is a trade-off between true positive and true negative rates; improving one often compromises the other.

To optimize the CDP, which encompasses Phases 1 through Approval as a whole, a stochastic decision process (SDP) can be used to determine the optimal policy that maximizes success or expected net present value (NPV) under practical constraints such as cost, timing, regulatory requirements, operational challenges, and competition. Actions in this context refer to clinical trial design, including design type (classical, adaptive, etc.) and go/no-go decision rules.

An SDP extends stochastic processes like Markov chains by incorporating decisions a_i, associated costs C_i, transition probabilities $p_{i,i+1}$, and gains g_{i+1} upon reaching the next state (Figure 5.5). The goal is to optimize actions (policies) to maximize expected rewards, often achieved using the backward induction method, based on Bellman's Optimality Principle:

"An optimal policy ensures that remaining decisions constitute an optimal strategy for the resulting state."

The expected reward at stage i with action a_i is:

$$g_i = p_{i,i+1}g_{i+1} - C_i.$$

Backinduction Calculations of Markov Decision Process for Clinical Development Program
(from g_4, g_3, g_2, to g_1)

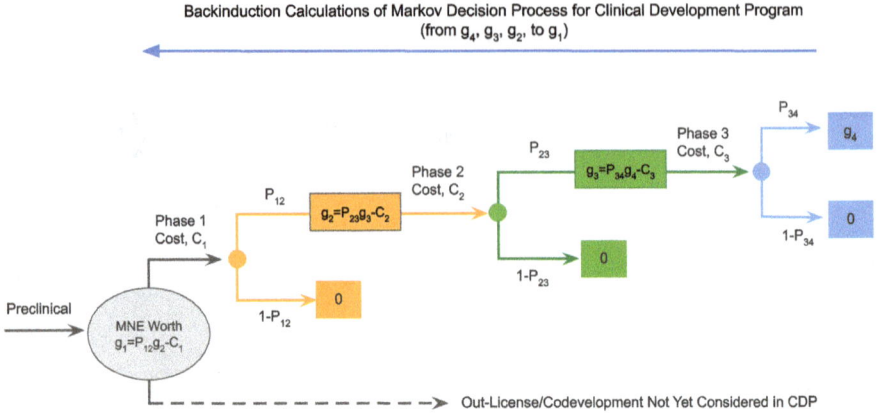

FIGURE 5.5
Illustration of MDP for CDP.

5.4.1 Key Considerations

1. Trial design: Operational costs (C_i) and transition probabilities ($p_{i,i+1}$) depend on the drug's safety and efficacy profile, trial design, and dose regimen. Different designs lead to varying recommendations and outcomes.

2. Monetary adjustments: Future rewards and costs are adjusted for inflation using a discount factor $(1+r)^t$, where r is the discount rate and t is the time to market.

3. Recursive calculation: Final reward (g_4) represents the drug's market value upon approval. Earlier rewards (g_3, g_2, g_1) are calculated recursively from g_4 without additional inflation adjustments, as intermediate gains are realized only through g_4 (Figure 5.5).

5.4.2 Example: SDP for a Cancer CDP

Using public data from an oncology indication, we construct an SDP by incorporating action options and corresponding gains at each milestone. Simulations highlight the impact of classical and adaptive designs on transition probabilities, success rates, and overall NPV.

Suppose the time value of a cancer drug candidate at approval (g_4) is $1,000 million, as estimated by the biotech company's commercial group. Using the phase transition probabilities from Figure 5.3 and assumed costs of $22.1 million, $11.2 million, and $3.4 million for Phases 3, 2, and 1 respectively, the expected NPV at each decision point is calculated through backward induction (Figure 5.6).

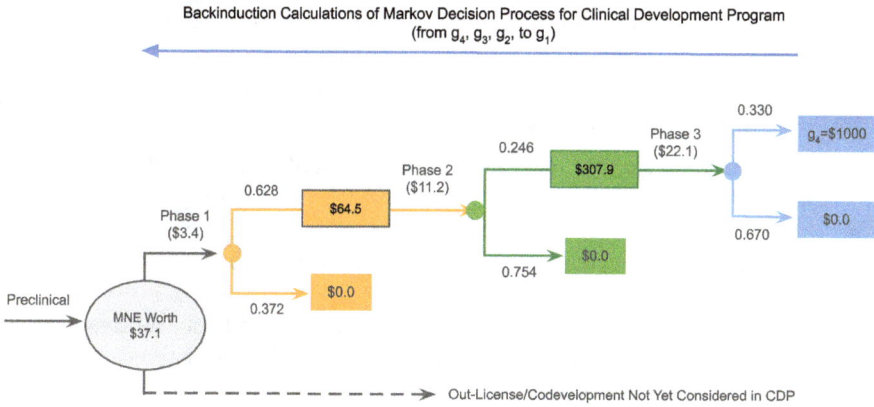

Backinduction Calculations of Markov Decision Process for Clinical Development Program
(from g_4, g_3, g_2, to g_1)

FIGURE 5.6
MDP for a cancer CDP.

1. Phase 3:
   - Cost: $22.1 million
   - Probability of success: 0.33
   - Expected value: $g_3 = (0.33 \times 1{,}000) + (0.67 \times 0.0) - 22.1$ = 307.9 million.

2. Phase 2:
   - Cost: $11.2 million
   - Probability of moving to Phase 3: 0.246
   - Expected value: $g_2 = (0.246 \times 307.9) + (0.754 \times 0.0) - 11.2$ = 64.5 million.

3. Phase 1:
   - Cost: $3.4 million
   - Probability of moving to Phase 2: 0.628
   - Expected value: $g_1 = (0.628 \times 64.5) + (0.372 \times 0.0) - 3.4 = 37.1$ million

In practice, trial designs differ in success probabilities, timelines, and costs. The optimization goal is to select the design that maximizes the NPV (g_i) at each decision point.

5.4.3 Q-Learning and Bayesian SDP

So far, we have only discussed a simple stochastic process, specifically the Markov decision process (MDP), where decisions depend solely on information from the previous stage. However, there are possible extensions to the MDP framework:

1. The reward may depend on both the state and the action, expressed as $g(s,a)$.
2. The reward may depend on the resulting state as well as the current state and action, expressed as $g(s,a,s')$.
3. The action space may vary by state, represented as A_s rather than a fixed A.

These extensions add complexity to the notation but do not fundamentally alter the problem or its solution method.

The MDPs discussed so far assume known transition probabilities, allowing calculations to be performed. However, what happens when these probabilities are unknown or overly complex, making them intractable? For example, in clinical development scenarios, the transition probabilities—and thus the expected overall gain—are significantly influenced by two unknown parameters: the toxicity rate and the normalized treatment effect. Changes in the assumptions about these parameters will alter the expected gain and, consequently, the optimal action rules.

To address this issue, methods such as Q-learning and the Bayesian SDP (BSDP) can be employed. For further details, refer to Chapter 5 of *Monte Carlo Simulation for the Pharmaceutical Industry* (Chang, 2011a).

5.5 Pharmaceutical Partnership: Shared Risk and Reward

5.5.1 Drug Development as Non-cooperative Game

In addition to making go/no-go decisions during clinical trials, a pharmaceutical company may choose to license out or partner with other organizations at a specific stage of the CDP.

Suppose two companies are deciding whether to develop competitive drugs for anemia, a condition characterized by a decrease in the normal number of red blood cells (RBCs) or a lower-than-normal quantity of hemoglobin in the blood. Since hemoglobin, found inside RBCs, carries oxygen from the lungs to the tissues, anemia's primary symptoms include general fatigue, weakness, malaise, and occasionally poor concentration.

The cost (including opportunity loss) of developing such a drug is denoted as C_A for company A and C_B for company B. The probability of successfully bringing the drug to the market is p. If a drug receives regulatory approval, the gross profit for a successful company depends on whether the competitor also succeeds. Assume the total market profit is fixed at G billion dollars if at least one company successfully develops the drug. If both companies succeed, they each receive $G/2$ billion dollars.

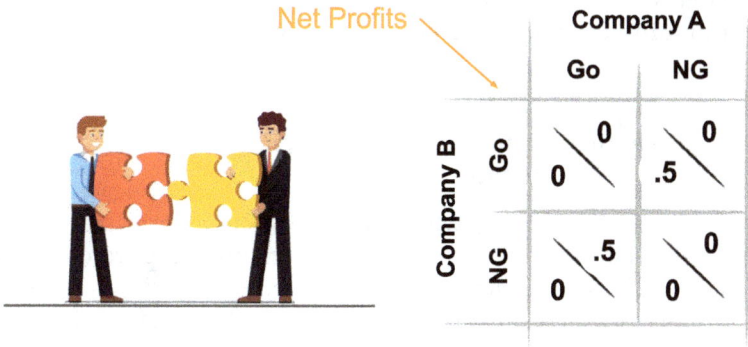

FIGURE 5.7
Pharmaceutical partnership game framework.

Given these conditions, the expected net profits are as follows:

- If both companies proceed with development (Go decision), their expected net profits are $pG/2 - C_A$ for company A and $pG/2 - C_B$ for company B.
- If only one company develops the drug while the other opts out (No-Go decision), the developing company earns $pG - C_A$ or $pG - C_B$, while the non-participating company earns zero.
- Given $p = 0.1$, $G = \$10B$, and $C_A = C_B = \$0.5B$, the net profits for all possible scenarios are summarized in Figure 5.7.

Here, we assume that when two companies share the market, savings in drug manufacturing offset the increased advertising costs due to competition. These figures in the example align with pharmaceutical industry benchmarks and are supported by two key studies. A 2020 study published in *JAMA* (Wouters et al., 2020) estimated a mean R&D cost of $1.3 billion per drug, accounting for clinical trial failures and opportunity costs. In contrast, the $0.5 billion cost cited here reflects the impact of cost-cutting measures, including early termination of unpromising trials, as well as reductions in marketing and manufacturing expenses.

The {Go, Go} scenario is a Nash equilibrium because neither company can improve their position by changing their decision unilaterally. If Company A decides to stop digging (choosing No-Go), it earns nothing while Company B gains the full profit ($pG - C_B$). Similarly, if Company B opts for No-Go, Company A benefits fully. Thus, both companies are locked into the Go decision to avoid losing out entirely.

However, this equilibrium illustrates a broader inefficiency. Both companies engaging in digging results in shared costs and diminished profits. This is analogous to a "tragedy of the commons" where individual players pursuing their self-interest reduce the overall value of the shared resource.

5.5.2 Pharmaceutical Partnerships: A Strategic Response to Competitive Dynamics

In the pharmaceutical game, where firms face high costs and uncertain rewards, partnerships offer a cooperative alternative to pure competition. Much like two treasure hunters pooling resources, partnerships enable companies to share risks, access complementary expertise, and accelerate innovation. These alliances span biotech firms, contract research organizations, academics, policymakers, and governments, taking forms such as spot market exchanges, contractual agreements, and mergers or acquisitions.

Alliances are collaborations where independent organizations pool resources while retaining strategic autonomy. These partnerships provide quicker access to skills, reduce costs, and foster flexibility. For example, firms can share clinical trial expenses or expedite regulatory compliance. However, risks exist, including adverse selection (partners misrepresenting abilities), moral hazards (contributing less than promised), and holdup (exploiting specific investments).

Cooperative strategies take two forms: collusion, where firms reduce industry competitiveness (e.g., price-fixing) and strategic alliances, fostering innovation without undermining competition. Examples include outsourcing tasks, licensing intellectual property, equity alliances, and joint ventures. Success depends on complementary skills and aligned goals, ensuring mutual benefit.

When deeper integration is required, firms may merge or acquire others. A merger combines two companies into a single entity, pooling resources for a competitive advantage. An acquisition involves one firm absorbing another, leveraging its assets or competencies, while a takeover occurs without the target's consent. However, over 50% of mergers and acquisitions fail due to cultural clashes, overestimated synergies, or poor integration.

In conclusion, partnerships and alliances offer an efficient alternative to the inefficiencies of competition, enabling firms to share costs and risks while innovating faster. Whether through outsourcing, licensing, joint ventures, or mergers, collaboration maximizes collective value, echoing the lessons of the pharmaceutical game: working together often achieves better outcomes than going it alone. For mixed n-player pharmaceutical game and expected payoff for mixed n-player game, see Chang's book: Monte Carlo Simulations for the Pharmaceutical Industry (Chang, 2011a; Chang, 2019).

5.6 Summary: Advancing Drug Development through Game Theory and Stochastic Processes

This chapter explored how concepts from cooperative game theory and stochastic processes can revolutionize clinical trial design and the broader CDP. By integrating these mathematical frameworks, the chapter addressed the challenges of optimizing decision-making under constraints like cost, time, and regulatory requirements, while managing uncertainties inherent to drug development.

Key Takeaways:

1. Game theory for strategic insights: Foundational principles such as Nash equilibrium, Pareto efficiency, and the prisoner's dilemma were used to illuminate how individual and collective decision-making can shape clinical trial outcomes. These concepts provided a basis for understanding trade-offs and designing strategies that balance competing endpoints like safety, efficacy, and cost.

2. SDPs for optimization: MDPs were extended into SDPs, incorporating flexible action spaces, state-dependent rewards, and strategies to handle unknown transition probabilities. This structured framework supports long-term optimization across all phases of drug development.

3. Innovative tools to address uncertainty: Techniques such as Q-learning and BSDPs were highlighted as powerful solutions for navigating complex and uncertain environments. These tools enable adaptive decision-making and optimize expected outcomes in dynamic clinical scenarios.

By bridging cooperative game theory and stochastic processes with clinical research, this chapter offers a forward-looking perspective on drug development. These methodologies not only improve efficiency and outcomes but also provide a robust framework for managing the uncertainties and complexities of modern pharmaceutical innovation.

Exercises

1. What critical thinking ideas and elements of creative analogies stand out to you in this chapter?

2. Explain how Nash equilibrium applies to decision-making in clinical trials. Use an example involving two pharmaceutical companies deciding on whether to launch competing drugs in the same market.

3. Construct a simplified SDP for a CDP. Define states, actions, transition probabilities, and rewards. Discuss how different assumptions about costs and probabilities affect the optimal strategy.

4. Describe how Q-learning can be applied to optimize clinical trial decision-making when transition probabilities are unknown. Provide a hypothetical example involving early-phase trials.

5. Define Pareto efficiency in the context of multi-objective optimization for clinical trials. Propose a trial design scenario where balancing safety and efficacy is required and discuss how Pareto efficiency could be achieved.

6. Analyze a scenario where two pharmaceutical companies collaborate on a clinical trial. Use concepts like shared risk and reward to propose an optimal partnership model.

7. Based on the chapter's discussion of pharmaceutical partnerships, design a hypothetical collaboration between two companies. Address key considerations like cost-sharing, intellectual property, and risk management.

8. Create a framework for a clinical trial that incorporates SDPs and game theory principles. Explain how your design improves upon traditional approaches.

9. Debate whether cooperative strategies (e.g., partnerships) or non-cooperative strategies (e.g., independent trials) are better suited for pharmaceutical companies. Consider factors like cost, risk, and time to market.

10. Propose a partnership model for two pharmaceutical companies that minimizes risks while maximizing collective rewards. How can concepts like shared costs and aligned incentives improve trial outcomes?

6

Enhancing Productivity and Economic Growth through Paradoxical Insights

Self-interested behavior can lead to the consequence no one wants.

When options are uneven, randomization is the way to achieve fairness.

6.1 Braess's Traffic Paradox: When Adding More Creates Less

Braess's paradox is the counterintuitive observation that adding one or more roads to a road network can worsen overall traffic flow rather than improve it. It suggests that increasing the number of routes available does not always reduce congestion and can sometimes lead to longer travel times for all drivers. This paradox was formally articulated by Dietrich Braess (1968), who demonstrated it with a specific road network model.

Braess's paradox has been observed in several real-world cases where altering road networks resulted in unexpected traffic behavior. Here are a few notable examples:

A famous example occurred when Seoul's Cheonggyecheon highway was removed as part of an urban renewal project. This elevated highway, which carried significant traffic through the city center, was demolished and replaced with a public park and stream. Surprisingly, the removal led to reduced congestion in the surrounding areas, as drivers took alternative routes that distributed traffic more efficiently across the network.

During the 1990s, several lanes on 42nd Street in Manhattan were closed for a parade, and officials anticipated significant traffic disruptions. However, traffic flow actually improved, with shorter travel times for many drivers. The temporary road closure forced vehicles to redistribute more efficiently across the available network, thereby alleviating congestion in some areas.

The paradox occurs because when new roads are added, individual drivers may change their routes to what seems optimal for them, but these decisions can lead to suboptimal outcomes for the entire network. This is a classic example of self-interested behavior leading to a worse collective outcome, where drivers seek to minimize their own travel time, inadvertently increasing overall congestion.

DOI: 10.1201/9781003630081-6

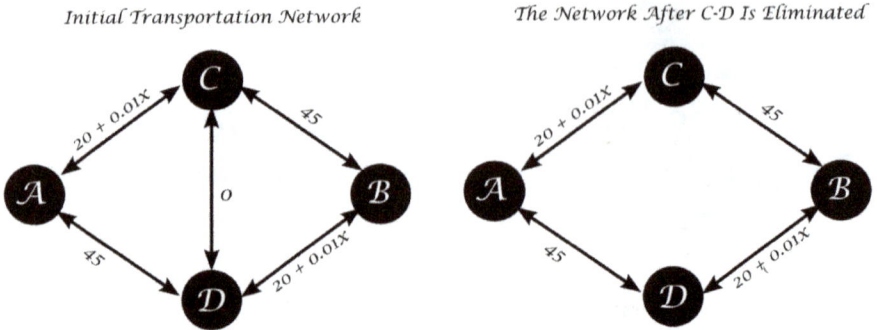

FIGURE 6.1
Braess's traffic paradox illustrated.

The concept has analogies beyond traffic flow, extending to electrical power grids, where adding transmission lines can increase energy losses, and biological systems, where introducing additional pathways or feedback loops can disrupt efficient functioning. In these cases, the structure of the network and the interactions between components can create unexpected effects.

Let's illustrate Braess's paradox using a simple numerical example with a road network (Figure 6.1).

Suppose that 2000 cars want to travel from A to B (the left Figure) over a weekend. A → C and D → B are two narrow roads, the travel time for each road increasing with the number of cars on the road. The travel time for n cars traveling on is $20+0.01n$ minutes. A → D and C → B are two wide roads, the travel time here being 45 minutes independent of n. C → D is a very short wide road with negligible travel time. Because of this, C and D can be considered as one place. If a traveler takes road A → C, it will take him 40 minutes at the most, when every traveler takes path A → C, which still is less than 45 minutes when traveling on path A → D. Therefore, everyone will take A → C. Similarly, at C, everyone will spend 40 minutes on path C → D → B and no one will take path C → B. Therefore, the total travel time from A to B will be 80 minutes for every vehicle. This is a self-interested behavior in road networks.

Now suppose the road C → D is removed by government intervention, leaving only two possible paths from A to B: A → C → B and A → D → B. At the Nash equilibrium, 1,000 travelers will take each of the paths. The travel time for each car is $45+20+0.01(1,000)=75$ minutes. This result is so surprising because the removal of the optional road C → D actually reduces everyone's travel time by 5 minutes or 6.25%. Even more surprising is that a budget cut (in maintaining the road C → D) by the government can actually do something good for everyone!

In a competitive market, companies or individuals might pursue strategies that seem optimal for them, such as increasing production capacity, expanding services, or lowering prices to gain market share. However, if all competitors adopt similar strategies, the result can be an oversupply of goods or services, driving down profits for everyone.

This is akin to adding a new road in the traffic scenario: while each company believes that expanding production will improve their own outcomes, the collective result can be an inefficient allocation of resources, lower prices, and reduced profits across the board.

In genetic networks, certain genes regulate the expression of others. Adding a new regulatory gene (e.g., through genetic modification) may appear to enhance the system's ability to respond to changes. However, this additional regulation can lead to unexpected feedback loops or cross-talk between pathways, resulting in inefficient gene expression or even maladaptive responses. This mirrors the paradox in traffic networks, where a new road complicates the flow rather than improving it.

In the circulatory system, adding new blood vessels (angiogenesis) is often thought to improve blood flow to tissues. However, in some cases, such as in tumor growth, the new blood vessels may be poorly organized, leading to inefficient blood flow and even hypoxic regions within the tissue. This example aligns with Braess's paradox, where adding more pathways (blood vessels) does not necessarily improve the overall function (blood supply).

In the brain, synaptic plasticity allows for the formation of new connections between neurons, which is essential for learning and memory. However, adding too many synaptic connections or hyperactive networks can lead to overexcitation and inefficiencies in signal transmission. For example, in conditions like epilepsy, the formation of excessive synaptic connections or abnormal neural pathways can result in uncontrolled neural activity (seizures), which reduces the efficiency of information processing rather than enhancing it.

Cells often have multiple pathways for processing nutrients and producing energy. If a new metabolic pathway becomes active due to mutations or environmental changes, it may divert resources away from other essential processes, leading to metabolic imbalances. For instance, in certain cancer cells, metabolic reprogramming adds additional pathways for energy production, but this can create bottlenecks in cellular metabolism and accumulation of toxic byproducts. This is similar to how adding a new road can create unexpected traffic problems.

The heart's electrical system coordinates contraction to pump blood effectively. If there are extra electrical pathways (such as in conditions like Wolff–Parkinson–White syndrome), they can cause abnormal heart rhythms (arrhythmias) instead of improving the heart function. In this case, the addition of extra pathways leads to less efficient pumping of blood, similar to how adding a new road can worsen traffic flow.

6.2 Boosting Workplace Efficiency through Counterintuitive Strategies

We now explore three enlightening examples that draw from the principles of Braess's paradox, illustrating how counterintuitive strategies can improve efficiency and outcomes.

Example 6.1: Optimizing Printing and Binding Costs

Consider a publishing company equipped with two types of printers and binding machines: small and large. The cost per book for the large printer is fixed at $10. For the small printer, costs increase with daily volume n due to inefficiencies, calculated as $2 + 0.07n$. Similarly, the cost per book for the large binding machine is $10, while the small binding machine follows the same formula, $2 + 0.07n$. The company produces 100 books daily across various departments, and each department seeks to minimize costs, as savings contribute to their annual bonuses (Figure 6.2).

Initially, all departments opt for the small printer and small binding machine. Their reasoning is straightforward: the large machines cost $10 per book, while the worst-case cost for the small machines is

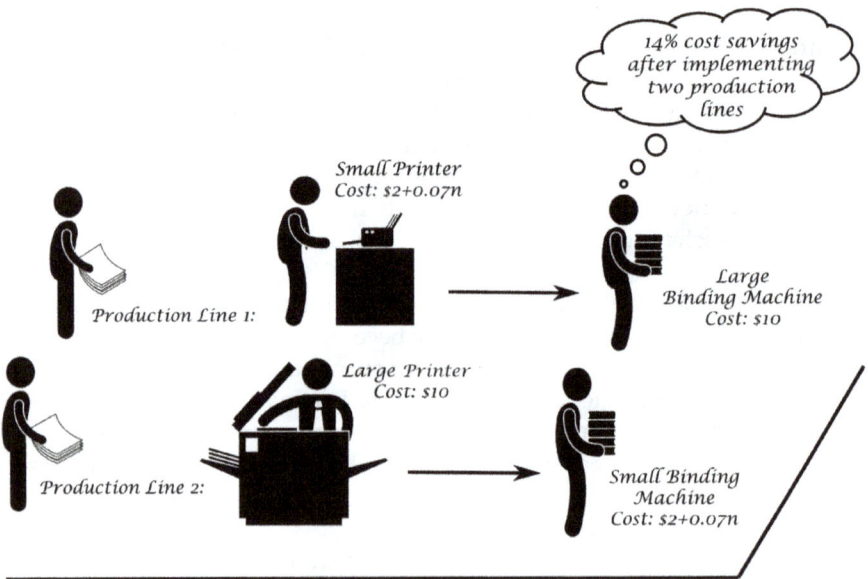

An Insight Derived from Braess's Paradox

FIGURE 6.2
Productivity gains inspired by Braess's paradox.

2+0.07(100)=9 per book. Combining printing and binding costs, they conclude that each book will cost at most $18. Consequently, producing 100 books will cost the company $1,800 daily.

However, insights from Braess's paradox reveal a more efficient solution. The company can establish two production lines:

1. A small printer paired with a large binding machine.
2. A large printer paired with a small binding machine.

If departments rationally distribute their workload evenly between these lines (Nash equilibrium), the cost per book decreases: $10+2+0.07(50)=\$15.50$ per book.

For 100 books, the daily cost becomes $1,550—a savings of $250 (14%) compared to the original $1,800. Additionally, this arrangement reduces production time compared to relying solely on small machines.

Example 6.2: Reducing Traffic Congestion

In large cities, motorists frequently complain about worsening traffic conditions. While public transportation is promoted as a solution, many drivers find it more convenient to drive. Suppose taking public transportation on a particular route takes 49 minutes. Driving n cars on the same route results in travel times of $25 + 0.001n$ minutes. With 20,000 car owners, if everyone drives, the travel time averages: $25 + 0.001(20,000) = 45$ minutes, which is faster than the bus. Consequently, everyone prefers driving.

To address congestion, the city government can implement a policy: cars with even-numbered license plates can only be driven on even calendar days and odd-numbered plates on odd days. This reduces the number of cars on the road to 10,000 daily, with the remaining drivers switching to public transit. The new average travel time becomes: $(25 + 0.001(10,000) + 49)/2 = 42$ minutes.

This policy saves each person 3 min/day while also benefiting the environment. Based on the 2024 U.S. GDP and gas costs, this time savings translates to an economic value of $6 per person daily. While this is a simplified example, it illustrates how such interventions can help balance traffic loads and enhance overall efficiency.

Example 6.3: Cooperative Farming for Optimal Profit

Farmers often face decisions about which crops or livestock to produce for maximum profit. Suppose raising cows yields a profit of $0.8-0.3p$, where p is the proportion of farmers raising cows. The alternative farming option yields a profit of $0.48-0.05(1-p)$, where $1-p$ is the proportion of farmers pursuing this alternative.

Without cooperation, the profit from cow farming is always higher. For $p=0.8-0.3(1)=0.50$. The alternate option yields a maximum profit of $0.48-0.05(0)=0.48$.

As a result, all farmers choose cow farming. However, this is suboptimal for maximizing overall profits. If the government introduces some incentive program to balance farming proportions, targeting $p=0.6$, the average profit becomes:

$$(0.8-0.3(0.6))(0.6)+(0.48-0.05(0.4))(0.4)=0.556.$$

This cooperative strategy achieves higher average profits for all farmers if the incentive ensures a somewhat even distribution of total profits or if the cow raiser is randomly chosen among applicants to maintain fairness. Maintaining this balance requires collaboration among farmers and government intervention to ensure a fair distribution of effort between the two types of farming.

In conclusion, the examples of publishing, traffic management, and farming highlight the power of counterintuitive strategies inspired by Braess's traffic paradox. By challenging conventional decision-making and encouraging cooperation, significant improvements in efficiency, cost savings, and resource allocation can be achieved across diverse domains.

6.3 Recasting Braess's Paradox into Mathematical Optimization

When fairness is ensured, such as through randomization, optimizing the collective gain becomes equivalent to optimizing the average gain for individuals. This allows the collaboration game inherent in Braess's paradox to be framed as an optimization problem.

Denoting individual gain (lost) functions under different strategies by $f(x)$ and $g(x)$ for $0 \le x \le n$, the goal is to maximize or minimize the objective function:

$$\min_{0 \le m \le n} \{C(m) = mf(m)+(n-m)g(n-m)\}.$$

Given the non-linear and potentially complex nature of $f(x)$ and $g(x)$, the optimal solution may not always be derived analytically. However, numerical methods provide a practical and reliable way to find the optimal m. Modern computational power enables exhaustive searches across all possible values of m, ensuring an accurate solution within a reasonable timeframe. The optimization algorithms are specified as follows:

1. Specify n, $f(x)$, and $g(x)$ for the given range $0 \le x \le n$.
2. Compute $C(m)$ for all integer values of m in the range $0 \le m \le n$
3. Identify the value of m that maximizes (minimizes) $C(m)$.

Recasting Braess's paradox as an optimization problem highlights the interplay between fairness and efficiency. By leveraging numerical methods, the solution becomes both accessible and robust, enabling practical application in resource allocation, traffic systems, and beyond.

6.3.1 Product Manufacturing: Racing against the Clock

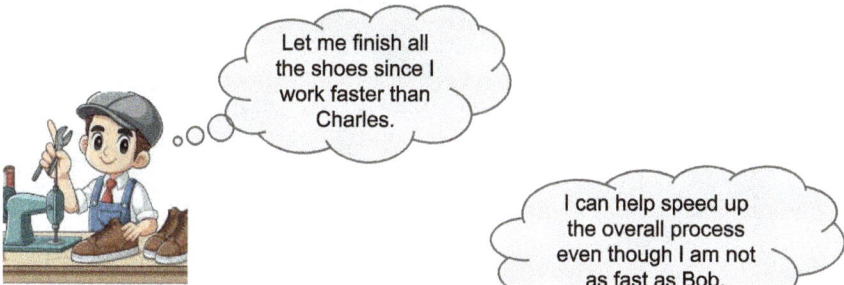

A small shoe factory operates with a single machine for producing shoes. Bob is an exceptionally efficient worker who takes pride in his work. On average, he can consistently produce shoes faster on average than Charles, even when working continuously for 10 hours. However, Bob's productivity peaks when working for shorter durations. Given that only one machine can be used by one person at a time, if your sole objective is to deliver the shoes as quickly as possible under time pressure, would you assign all the work to Bob, or would you also involve Charles to share workload (Figure 6.3)?

Let us reform the scenarios for clarity:

1. Bob's productivity curve:
   - Bob works faster than Charles on average over time, but his efficiency declines when working continuously for extended periods.

a)   One Person Works

b)   Two People Sequentially Work

FIGURE 6.3
Sequential collaboration: two workers outperforming one.

- There is a "diminishing returns" effect: as Bob works longer, his speed per pair decreases.
2. Charles's consistent productivity:
    - Charles is slower than Bob but maintains a consistent pace over time.
3. Time pressure:
    - The goal is to deliver shoes as quickly as possible. In this case, you need to balance maximizing immediate productivity with preventing burnout or inefficiency.
4. Only one person can work in the production line.

To meet the client's requirement as soon as possible and since Bob always works faster than Charles, the expected answer might be: Assign the entire task to Bob while staying within ethical limits to ensure faster overall delivery.
   However, from Braess's paradox, we know this apparent right answer could be wrong, as shown in the following numerical example (Figure 6.4).

- Total $n=100$ pairs of shoes need to be manufactured as soon as possible.
- Bob can produce a pair in $f(m) = 1 + 0.05m$ minutes on average, that is, Bob's efficiency reduces as the number of pairs (m) he needs to produce increases.

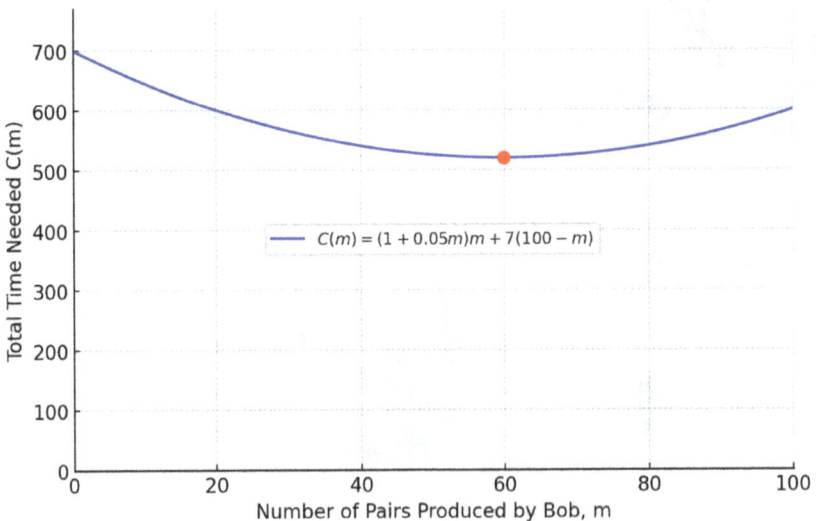

FIGURE 6.4
Shoe-making time versus work distribution.

- Charles needs $g(m) = 7$ minutes to produce a pairs, independent the number pairs $(100-m)$ he produces,

We need to minimize the time of manufacturing:

$$min_{0 \leq m \leq 100} \{C(m) = (1+0.05m)m + 7(100-m)\}.$$

Compare the values of $C(m)$ at the critical point and endpoints:

- $C(0) = 700$ (7 min/pair)
- $C(60) = 520$ (with derivative $\dfrac{dC(m)}{dm} = 0$)
- $C(100) = 600$ (6 min/pair, faster than Charles)

The minimum value of $C(m)$ is 520, and it occurs at $m=60$.

The key insight here is that Bob remains more productive than Charles even when working for a long time. However, Bob becomes even more efficient when Charles takes on some of the work and Bob works for a shorter duration. The time required for Bob to make the 61st pair of shoes is 7.05 minutes, which is slightly more than the 7 minutes needed for Charles.

This is a hypothetical example, but scenarios resembling Braess's Paradox can arise in real-world production and logistics when choosing between two production lines or routes—one with fixed efficiency and the other with efficiency that varies based on production volume due to emerging bottlenecks.

6.3.2 Optimizing Drug Effectiveness through Target Population Strategies

The concept can be likened to Braess's Paradox, where the effectiveness of a treatment for an individual is not constant but depends on the size and diversity of the population receiving the treatment. For example, expanding the target population for a drug to include a less effective patient group without proper stratification can reduce its overall effectiveness, much like traffic congestion increases when additional routes are introduced without optimizing flow.

While biomarkers are commonly used to stratify patients and identify subgroups with higher responsiveness, these tools may be unavailable, invasive, or prohibitively expensive. Moreover, treating all patients without pre-screening may seem more equitable but risks diluting the overall effectiveness of the therapy.

When a treatment is applied to a broader, heterogeneous population, several factors contribute to reduced effectiveness:

- Diverse patient profiles: Variability in patients' biology and disease characteristics makes it challenging to tailor doses, leading to less precise treatment regimens and diminished outcomes.

- **Reduced dosing precision**: To ensure safety for a larger population, lower doses may be used, resulting in weaker effects.
- **Biomarker screening limitations**: While biomarkers can differentiate treatment effects and optimize targeting, practical limitations such as unavailability, invasiveness, or high costs of screening tools often prevent their use. Additionally, excluding certain patients based on biomarker status may seem unfair in diseases like cancer, where patients seek access to any potential treatment.

In real life, there are some drugs that have the characteristic differential effectiveness for different populations and alternative drugs have different efficacies and safety profiles, as shown below:

Pembrolizumab (Keytruda) for Various Cancers

- **Differential effectiveness**: Patients with low PD-L1 expression or microsatellite stable tumors exhibit reduced response rates. Biomarkers such as PD-L1 expression help predict who will benefit most.
- **Alternatives**: Other immune checkpoint inhibitors, such as Nivolumab (Opdivo), target similar pathways (PD-1, PD-L1, or CTLA-4) to enhance the immune system's ability to fight cancer cells.

Warfarin (Coumadin) for Blood Clot Prevention

- **Differential effectiveness**: Patients with different genetic profiles show varied responses to the same dose of warfarin, affecting both safety and effectiveness.
- **Alternatives**: Direct oral anticoagulants (DOACs) or novel oral anticoagulants (NOACs) offer improved convenience and safety compared to Warfarin, making them popular choices in modern practice.

To improve outcomes, strategies must balance effectiveness with accessibility:

- **Narrowing the target population**: Focus on patients most likely to benefit based on biomarkers or other predictive factors.
- **Improving biomarker access**: Invest in less invasive, more affordable screening tools to make biomarker-driven treatment more practical.
- **Broad population equity**: Develop policies or protocols to ensure fairness when treatments are offered to broader populations, even if some patients may benefit less.

By refining patient targeting and leveraging biomarkers where feasible, healthcare systems can optimize drug effectiveness while minimizing unintended reductions in efficacy when broad populations are treated. This approach not only aligns with scientific evidence but also supports ethical and equitable care.

We now can formulate the optimization for the problem.

Drug response depends on many factors, with the biomarker being just one of the identified determinants. Assume the effectiveness of a treatment is a function of the target population composition:

- p: The estimated proportion of biomarker-positive patients in the population based on screening results.
- m: The number of total biomarker-screened patients for treatment in the real world, not just in clinical trials.
- a: The effectiveness of treating screened biomarker-positive patients, with no significant gain in the benefit-risk ratio for treating biomarker-negative patients. Note: The treatment effect already accounts for the smaller effect of false-positive patients.
- b: The effectiveness for the overall patient population without screening for the biomarker, where the total patient population size is n. The dose may be adjusted for treating different populations.

If we screen for the biomarker, we can only treat biomarker-positive patients because our assumption is that the lower benefit-cost ratio does not justify treating biomarker-negative patients. If we do not screen for the biomarker, the drug can be used for the overall patient population. A third approach combines the two strategies, which leads to the optimization in terms of total benefits of the treated population:

$$\max_{0 \leq m \leq n} \{U(m) = mpa + (n - m)b\}, \quad \text{where} \quad a > b > 0.$$

Here, mpa is the total treatment benefits for screened biomark-positive patients, and $(n-m)b$ is the total treatment benefits for the unscreened patients.

The parameters $p, a, b,$ and m can be determined through biomarker-enabled clinical trials. Other factors may also need to be considered including cost of, inconvenience due to, and invasion caused by screening, and cost of the drug.

Since $U(m)$ is a linear function of m, the optimal solution is:

1. Full screening $(m = n)$: If $p \cdot a > b$, screening the entire population for biomarkers is optimal.
2. No screening $(m = 0)$: If $p \cdot a \leq b$, no screening is optimal.

This framework allows for the systematic evaluation of the trade-offs between targeted screening and broad treatment approaches to maximize patient benefit while accounting for practical constraints.

The optimization can be extended to more options $g_i(x)$, to maximize utility function:

$$\max \sum_{i=1}^{m} x_i g_i(x_i), \quad \text{where} \quad \sum_{i=1}^{m} x_i = n.$$

The maximization is regarding x_i, $i = 1, \ldots, n$.

6.4 Harnessing Creative Analogies in Economic Problem-Solving

Economics is a field rich with creative analogies and critical thinking, as it seeks to explain complex systems and human behaviors using relatable comparisons, thought experiments, and simplified models. Here are some excellent examples:

6.4.1 Broken Window Fallacy

The Broken Window Fallacy, introduced by French economist Frédéric Bastiat in his 1850 essay *What is Seen and What is Not Seen*, challenges the idea that destruction fuels economic growth. In this scenario, a shopkeeper must pay a glazier to repair a broken window, seemingly creating work and stimulating spending. However, this does not generate new wealth—the shopkeeper could have used that money for other productive purposes, meaning the apparent economic gain is merely a reallocation of resources rather than true growth.

Bastiat highlights the flaw in this reasoning: it focuses only on what is seen—the visible economic activity of the repair—while ignoring what is unseen—the opportunity cost of the shopkeeper's lost spending power. Had the window not been broken, the shopkeeper might have invested in new tools, expanded inventory, or pursued other ventures that generate lasting value. In this way, destruction does not create wealth but merely redistributes existing resources, potentially leaving the economy worse off.

This fallacy underscores the need to consider both direct and indirect economic effects. While some Keynesian economists argue that repairing damages during a recession can temporarily boost demand and employment, Bastiat's insight emphasizes that this is a short-term fix. True economic growth stems from productive investments that build lasting value, not from shifting resources to replace losses. Modern parallels include disaster recovery efforts, which, while temporarily stimulating economic activity, do not account for the wealth lost and the human suffering caused by disasters. The broken window fallacy remains a timeless reminder to consider both the seen and unseen impacts of economic actions.

6.4.2 The Keynesian Paradox of Thrift

The paradox of thrift, a key concept in Keynesian economics, illustrates how rational individual behavior can lead to negative collective outcomes during economic downturns. When households increase their savings and cut spending, aggregate demand declines, reducing overall economic activity and, paradoxically, weakening the very savings they intended to strengthen.

This paradox reflects the broader tension between individual rationality and collective irrationality, a dynamic that fuels debates on fiscal policy and government intervention. Keynes argued that such situations necessitate countercyclical government action—increased public spending during recessions to counter reduced private consumption, stabilize aggregate demand, and promote recovery. While saving is essential for long-term investment, unchecked thrift during economic contractions can deepen recessions, highlighting the need for a balance between private prudence and public intervention.

This insight provides a compelling analogy: just as an individual saving more is rational, if everyone saves simultaneously during a downturn, the economy contracts, making it harder for individuals to achieve their financial goals. The paradox of thrift thus reinforces the importance of coordinated policy responses to bridge the gap between personal financial decisions and broader economic stability.

6.4.3 The Cobra Effect

A bounty on cobras incentivizes breeding rather than eradication, leading to an even larger cobra population when the program ends. This illustrates how poorly designed incentives can backfire, producing unintended consequences.

The cobra effect exemplifies this dynamic, where well-intentioned policies lead to perverse outcomes. During British colonial rule in India, a bounty on dead cobras aimed to reduce their population. However, enterprising individuals began breeding cobras to collect the rewards, ultimately undermining the policy's goal. When the bounty was discontinued, the breeders released the now-worthless cobras, increasing the population rather than reducing it.

This phenomenon highlights the principle that incentives matter, but misaligned incentives can encourage counterproductive behaviors. In workplaces, performance metrics can drive superficial success—such as prematurely closing customer tickets—rather than solving underlying problems. Similar patterns emerge in other fields: fee-for-service models in healthcare can promote unnecessary procedures, while performance-based funding in education may incentivize grade inflation or teaching to the test rather than genuine learning.

To avoid these pitfalls, policymakers must anticipate unintended consequences by designing incentives that align with long-term objectives. The cobra effect serves as a cautionary tale, emphasizing the need to evaluate not just the intended outcomes of a policy, but also the potential for perverse behaviors.

Taken together, the broken window fallacy, the paradox of thrift, and the cobra effect illustrate the complexities of economic decision-making. They reveal the tension between individual and collective rationality, the unseen consequences of actions, and the crucial role of well-designed incentives. These paradoxes challenge intuitive assumptions, deepen our understanding of economic systems, and encourage critical thinking about both short- and long-term consequences. Engaging with these ideas equips us to navigate modern economics with greater insight and effectiveness.

6.4.4 The Hotel Room Parable

This is a classic economic parable illustrating the circulation of money and how liquidity can temporarily resolve debts. Here is how the story goes:

A businessman arrives in a small town and walks into the only hotel. He approaches the hotel owner and says, "I need a suitable room for the night. Here's $1,000 as a deposit while I check the rooms."

The hotel owner takes the $1,000 and immediately runs to the butcher, to whom he owes money, and pays off his debt. The butcher then rushes to pay the farmer who supplies him with meat. The farmer, in turn, goes to settle his outstanding bill with the town's doctor. The doctor, who had an unpaid tab at the hotel, finally returns the $1,000 to the hotel owner.

Meanwhile, the businessman checks the rooms but finds none to his liking. He returns to the front desk, takes back his $1,000 deposit, and leaves the town without spending a dime.

Yet, in the span of just a few minutes, all debts in the town have been settled, and no actual money stayed behind.

This story highlights the power of money circulation in an economy. Liquidity, even if temporary, can resolve financial bottlenecks, reduce stress, and keep businesses running. It also shows that sometimes, a problem isn't a lack of wealth but a lack of movement in the system.

6.5 The Equal Opportunity Paradox

The idea of equal employment opportunities is often seen as a cornerstone of fairness and progress. By offering everyone an equal chance to succeed, it aims to reduce disparities and empower individuals, regardless of their background. However, when applied to rural populations, equal opportunities can lead to unexpected consequences. These include a phenomenon where the most capable individuals migrate to urban areas (developed countries), leaving rural communities (developing countries) further disadvantaged. This dynamic creates a paradox: while equal opportunities help individuals thrive, they can inadvertently deepen inequalities at the community level.

6.5.1 The Paradox

Equal employment opportunities, while seemingly fair, often have unintended consequences:

1. Selective outmigration: The "brain drain" effect occurs when equal opportunities favor individuals with higher skills, better education, or greater ambition. These individuals, often the most capable of driving local development in rural areas, leave, resulting in a loss of potential leaders, innovators, and contributors, exacerbating poverty and stagnation. Demographic shifts further complicate the issue, as the outmigration of younger, more educated individuals often leaves behind an aging population, limiting the rural area's ability to sustain growth.

2. Urban–rural divide: Equal opportunities are not truly equal when systemic barriers, such as lack of education, healthcare, and infrastructure, prevent rural individuals from competing effectively with their urban counterparts. This perpetuates inequality despite the system's nominal fairness.

3. Resource reallocation: Resources directed toward creating opportunities for rural individuals in urban areas may inadvertently reduce investments in rural regions themselves. This reallocation can lead to further impoverishment and underdevelopment of rural communities.

6.5.2 Addressing the Paradox

To mitigate the potential negative impacts of this paradox, a more nuanced approach to equal opportunities is required—one that balances individual advancement with community sustainability.

1. Promote local development by decentralizing opportunities. Job and education opportunities within rural areas should be created to reduce the need for outmigration. For example, investments in rural industries such as sustainable agriculture, renewable energy, or rural tourism can be prioritized, along with support for small and medium enterprises (SMEs). Improving infrastructure like roads, schools, hospitals, and internet connectivity is essential for attracting businesses and retaining talent in rural areas.

2. Encourage circular migration by incentivizing return migration. Providing grants or entrepreneurial support can encourage

individuals to return to rural areas after gaining education or work experience in urban centers. Hybrid work models and remote job opportunities can also allow rural residents to participate in urban economies while remaining in their communities.

3. Balance individual and collective benefits by implementing community-based approaches. Policies should ensure that the benefits of equal opportunities extend to rural communities. Urban employers could be required to invest in rural education or community development as part of their hiring practices. Cooperative models can be introduced, allowing individual successes to translate into broader benefits for rural hometowns.

4. Focus on capacity building through improved access to education and vocational training in rural areas. This ensures that opportunities are accessible to a broader segment of the rural population, not just the most skilled. Empowering local leaders who can drive sustainable development initiatives is also crucial.

5. Address systemic inequalities by reforming structural barriers that disadvantage rural communities. Policies should focus on reducing disparities in healthcare, legal protections, and public services. Equitable policy design must consider the unique challenges and needs of rural areas to ensure fairness.

This paradox reflects a broader tension between individual mobility and collective well-being. Systems that prioritize one over the other can lead to unintended consequences. Addressing the paradox requires shifting from a purely individual-centric model of equal opportunity to one that integrates community and systemic perspectives. By fostering environments where rural communities can thrive while allowing individuals to pursue their potential, the cycle of rural impoverishment can be broken, leading to a more equitable society.

6.6 Summary: Counterintuitive Insights for Economic Growth and Productivity

This chapter examines paradoxical insights, such as Braess's Paradox, to reveal how counterintuitive strategies can enhance productivity and economic systems. It demonstrates how adding constraints or rethinking resource allocation often yields better outcomes in diverse contexts, from traffic management to workplace efficiency and healthcare. The chapter also highlights economic paradoxes like the broken window fallacy and the paradox of thrift, which challenge conventional assumptions about growth and rationality.

6.6.1 Critical Thinking Perspective

1. Challenging assumptions: The chapter questions traditional beliefs that "more is better," demonstrating through paradoxical insights that over-optimization can lead to inefficiencies, whether in traffic networks, workplaces, or healthcare systems.

2. Analyzing systemic effects: By considering both direct and indirect consequences, the chapter emphasizes the need for holistic thinking in decision-making. For example, Braess's Paradox illustrates how individual self-interest can degrade collective outcomes.

3. Balancing trade-offs: The chapter urges readers to evaluate trade-offs between short-term gains and long-term sustainability, as seen in discussions of equal opportunity paradoxes in rural development.

6.6.2 Creative Analogy Perspective

1. Braess's paradox as a framework: Analogies drawn from traffic systems, such as how removing roads can improve traffic flow, creatively explain broader applications in resource allocation and economic optimization.

2. Economic fallacies as analogies: Concepts like the broken window fallacy and cobra effect highlight how poorly designed incentives and misallocated resources can backfire, using vivid, relatable examples to simplify complex economic dynamics.

3. Interdisciplinary connections: Analogies from biology, such as angiogenesis or synaptic plasticity, link the principles of Braess's Paradox to systems in healthcare, emphasizing the universality of these counterintuitive insights.

By combining critical thinking and creative analogies, the chapter reframes conventional wisdom, inspiring readers to explore counterintuitive solutions for improving productivity, fairness, and economic resilience across systems.

Exercise

1. Explain Braess's paradox using a real-world scenario (e.g., traffic management and workplace efficiency). Why does adding more options sometimes lead to worse outcomes?

2. How does randomization play a role when translate Braess's paradox problem into an optimization problem?

3. What critical thinking ideas and elements of creative analogies stand out to you in this chapter?

4. Discuss how Braess's paradox can apply to workplace productivity. Propose a hypothetical scenario in which reducing options or resources improves outcomes.

5. Explore the broken window fallacy or the cobra effect. Provide a modern-day example where these paradoxes can inform better decision-making in economics or policy.

6. Develop an analogy similar to Braess's paradox in a different domain, such as healthcare, education, or sports. Explain its relevance and implications.

7. The chapter discusses randomness as a way to achieve fairness. Propose a system (e.g., in hiring, education, or sports) where randomness could improve fairness or outcomes.

8. Debate whether counterintuitive strategies, such as removing resources or options, are sustainable solutions. Use examples like urban traffic management or team productivity.

9. Discuss the equal opportunity paradox. How can policies balance individual success with community well-being in education or employment?

10. Solve an optimization problem inspired by Braess's paradox. For example, calculate the optimal allocation of resources in a workplace with constraints on time and efficiency.

11. Design a system to allocate public goods (e.g., park access and healthcare) that incorporates randomness or paradoxical insights. Explain its advantages and challenges.

7

Identifying Biases via Critical Thinking

Bias is both the flaw and the feature of human cognition. A world without bias is as unimaginable as one without perspective. Recognizing bias is the first step toward genuine understanding.

Individually, we all hold biases, but within a defined system, collective unbiasedness emerges through its overarching criterion.

7.1 Observation and Data Collection

Confirmation bias refers to the tendency to search for, interpret, favor, and recall information in a way that confirms pre-existing beliefs or hypotheses. People do this because it is more psychologically comfortable to reinforce beliefs than to challenge them. This behavior can occur both intentionally and subconsciously, leading individuals to overlook or minimize evidence that contradicts their views, thus creating a biased understanding of reality (Figure 7.1).

7.1.1 Examples of Confirmation Bias Exist Everywhere, Anytime

In political views, a classic example of confirmation bias is seen in political discussions. People tend to follow news sources or social media accounts that align with their political views, consuming content that reinforces their beliefs and dismissing or ignoring opposing viewpoints.

In health decisions, someone who believes that a certain diet or supplement is beneficial might only seek out success stories or studies that affirm its effectiveness while ignoring studies that show no benefit or potential harm.

In scientific research, confirmation bias can manifest when researchers cherry-pick data that supports their hypothesis or give more weight to evidence that confirms their assumptions while downplaying or disregarding evidence that does not. For example, in drug trials, a researcher may focus on positive results from subgroups of patients where the drug appears effective, while ignoring data from other subgroups where the drug shows no effect or adverse outcomes. This selective reporting can lead to biased conclusions about the drug's efficacy.

DOI: 10.1201/9781003630081-7

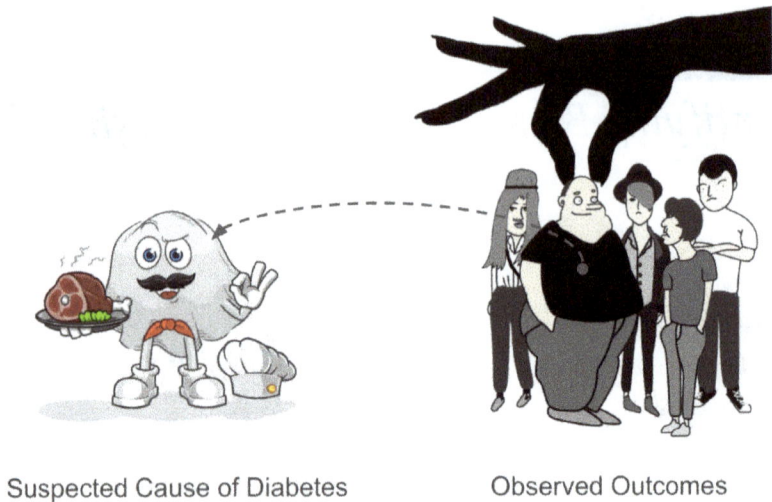

Suspected Cause of Diabetes   Observed Outcomes

FIGURE 7.1
Cherry-picking leading to biased observations.

Publication bias is a broader manifestation of confirmation bias in the scientific community. Studies with positive results are more likely to be published than those with negative or null results. This creates a skewed representation of research outcomes in the literature.

In contentious fields like climate science, confirmation bias is evident among both advocates and skeptics. Climate change advocates often highlight studies emphasizing severe impacts, while skeptics may focus on data that appears to downplay its significance. Skeptics might argue that extreme weather will naturally intensify over longer observation periods, suggesting that more extreme data points are simply expected with time.

This raises several critical questions: Have extreme climate events historically driven evolution? Should human activities be seen as part of natural evolution? And, are we giving equal attention to both extreme and non-extreme data, given their distinct impacts on society?

Heightened awareness fostered by advocacy can sometimes deepen social and climate issues, even though the extent of this effect is difficult to measure.

7.1.2 Common Strategies in Reducing Confirmation Bias

A crucial step in mitigating confirmation bias is developing awareness of its influence. Actively seeking out and engaging with opposing viewpoints—creating an analogy between yourself and the other side—enables a more balanced perspective. Critically evaluating evidence from all angles fosters deeper, more objective understanding.

In research, scientists must rigorously test their hypotheses by looking for disconfirming evidence as much as confirming evidence. The use of double-blind studies, randomization, and peer review helps minimize the risk of confirmation bias.

The requirement for replicability of study results is an effective way to reduce confirmation bias. The movement toward open science (where data and methodologies are shared transparently) also helps combat confirmation bias, as others can more easily scrutinize the research process.

It is interesting to know that, individually, we all hold biases, but within a defined system, collective unbiasedness emerges through its overarching criterion.

In the next sections, we will dive into some more bias issues and examine the technical details.

7.2 Regression to the Mean

The phenomenon of regression to the mean was first identified by Sir Francis Galton in the 19th century. Galton observed that sons of very tall fathers tended to be shorter than their fathers, while sons of very short fathers tended to be taller. This statistical tendency for extreme values to move closer to the average over time became known as regression to the mean.

A classic example of the regression fallacy can be seen in how improvement scores on standardized tests were calculated and interpreted in Massachusetts. In 1999, schools were assigned improvement goals, and the Department of Education measured changes in average student scores from 1999 to 2000. Interestingly, many of the lowest-performing schools met their improvement goals, while many of the highest-performing schools did not. However, the interpretation overlooked the phenomenon of regression to the mean: low-performing schools may have shown improvement partly because their scores, influenced by chance and temporary factors, tended to shift closer to the average in subsequent testing. Conversely, high-performing schools were more likely to see a natural decline.

Regression to the mean also explains why replicating others' success is difficult: Success often results from both skill and a degree of luck. Without the same "lucky" conditions, extreme success is unlikely to repeat.

Regression to the mean is a well-recognized phenomenon in clinical trials, particularly when participants are selected based on extreme baseline values. For example, a trial might enroll patients with hemoglobin levels below 10. Some of these patients will have lower-than-average hemoglobin levels at the start simply due to random variability, and their levels are likely to shift toward the average as the trial progresses, independent of any treatment. Consequently, the observed treatment effect in the study reflects not only the drug's actual impact and any placebo effect but also the influence of regression to the mean. Including a control group in clinical trials helps reduce this bias, allowing for a more accurate estimate of the treatment effect above and beyond the placebo effect.

7.3 Knowledge and Longevity: Competing Risks and Survivorship Bias

Everyone ultimately faces mortality, so a reduction in deaths from one disease will, in turn, shift mortality to other causes. Thus, an observed increase in deaths from one category does not necessarily indicate a decline in healthcare quality for that condition.

Figure 7.2 illustrates that deaths in the Cancer and Heart Disease categories (competing risks) decreased during 2020–2022, primarily due to the impact of the competing risk posed by COVID-19 rather than significant advancements in medical technology in these areas. A slight increase in unintentional injuries may be attributed to indirect effects of COVID-19, potentially increasing the likelihood of such incidents.

The reduction in deaths across all five categories by 2022 likely reflects the earlier mortality of vulnerable patients during the pandemic. Many individuals who succumbed to COVID-19 might have otherwise died in later years

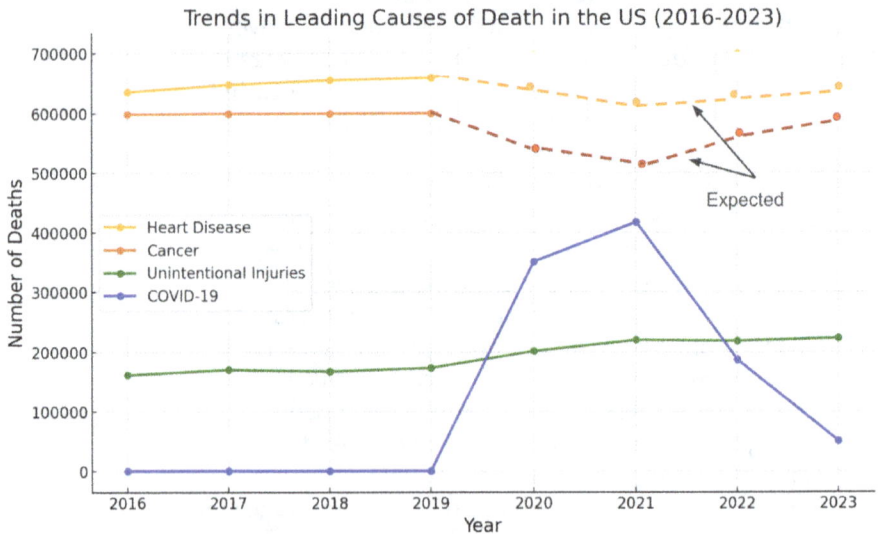

FIGURE 7.2
Competing risks illustrated through survival analysis.

without the pandemic. Consequently, it is expected that total deaths will decrease over the next 5–10 years, reflecting a longer-term "positive" effect of COVID-19 on mortality trends despite its devastating short-term impact.

This highlights the importance of assessing the overall impact of events like COVID-19 by considering both short- and long-term effects. Focusing on a single time frame risks producing a biased or incomplete understanding of the pandemic's full impact.

A similar example can be given: While Nobel laureates' life expectancy averages between 75 and 80 years (even accounting for historically shorter lifespans), the global average is approximately 72.6 years, according to the World Health Organization's 2023 data. From this, would you conclude that knowledge prolongs life? A key factor to consider is that achieving significant milestones—such as winning a Nobel Prize—often requires living long enough to reach them in the first place.

7.4 Funding Allocation and the Influence of Grouping Bias

The distribution of funds for disease areas by the U.S. government, particularly for research, is primarily determined by a combination of factors, including public health priorities, disease burden, scientific opportunity, advocacy, and legislative mandates.

However, when evaluating public health priorities and disease burden, conclusions heavily depend on categorization choices and thus could lead to bias. For instance, combining multiple diseases into a broader category can amplify perceived impact, while dividing COVID-19 into subcategories, as we sometimes do with rare diseases, and could make each COVID type appear to have minimal impact.

One might argue that COVID-19 deaths are relatively minimal compared to non-COVID deaths, questioning why it warranted such extensive measures and sacrifices in other areas. The counter-argument is that COVID-19 posed an immediate public health threat, with significant societal impact and health system disruptions. However, this reasoning risks circularity because "threat," "impact," and "disruption" depend on how we categorize and compare COVID-19, likely relative to other specific diseases in our minds.

Ultimately, the choice to group or split categories does not eliminate bias—it merely shifts the perspective. The key is to be transparent about how categorization influences conclusions, acknowledging that each approach will highlight different aspects of the problem.

Consequently, the distribution of funds based on disease impact and burden is significantly influenced by subjective disease categorization. How diseases are grouped or separated can shape perceived priorities and, consequently, funding allocations.

Furthermore, since each patient has an equal right to care, allocating more funding to cardiovascular disease for its higher cost-effectiveness can raise equity concerns, even if it does not intentionally disadvantage patients with other conditions, such as cancer. Balancing cost-effectiveness with a commitment to equity requires a nuanced approach that considers both the societal benefits of addressing high-impact diseases and the ethical responsibility to support all patients equitably.

If we fully recognize the right of every individual to identify themselves based on unique or self-recognized characteristics (such as gender identity, ethnicity, medical condition, etc.), achieving fairness becomes not only complex, but also impossible. Some people may like to simply categorize mankind into two categories: man and female, even though they fully recognize each individual is unique in their own way. Should those people sacrifice their freedom right and be forced to group people into more categories based on others criteria?

My point emphasizes a key aspect of individual freedom: just as people have the right to choose and express their unique identity, others also have the right to categorize based on their own criteria. This is not necessarily a denial of individual uniqueness but rather a way to organize information or understand groups, much like categorizing people by age. Just because we group individuals by age does not imply that everyone within that age group is the same in experience or outlook.

In other words, recognizing uniqueness and choosing categories are not mutually exclusive. People should be free to identify themselves in ways that feel authentic to them, and at the same time, others should have the freedom to categorize based on broad characteristics for practical or contextual reasons. This balance respects both individuality and the social or functional needs for categorization.

Ultimately, it is about allowing space for personal identity and contextual categorization to coexist without diminishing the value or freedom of either.

7.5 Publication and Funding Biases

7.5.1 Publication Bias

Publication bias refers to the tendency of research studies with positive, statistically significant, or novel results to be more likely published than those with negative, inconclusive, or less impactful findings. This bias arises from various systemic factors and has far-reaching consequences for the integrity of the scientific record.

Causes of publication biases include the following 4 categories:

1. Editorial preferences: Journals often prioritize studies with striking or groundbreaking findings, as these attract attention and citations. Negative or null results are perceived as less interesting.

2. Researcher behavior: Scientists may avoid submitting negative findings due to fears of rejection or reduced academic prestige.

3. Funding influences: Sponsors, including governments and private entities, may prefer results aligned with their priorities, indirectly shaping what gets published.

4. Statistical significance: Studies with significant results are seen as more compelling and are thus more likely to be published.

Publication bias related to statistical significance can manifest in two key ways. First, studies with statistically significant results are more likely to be published compared to those with non-significant or null findings. Second, among the studies that are published, the reported results often tend to be inflated rather than conservative or understated as discussed. In simpler terms, while some positive findings genuinely reflect strong effects, others may arise purely from random chance. This results in a skewed body of published research, disproportionately favoring positive or "good" results as discussed in *Regression to the Mean* with a numerical example.

For this reason, pharmaceutical companies, government regulatory agencies, and even individual researchers—despite having no direct conflicts of interest—may inadvertently introduce bias into their publications and drug labeling. This highlights the systemic challenge of avoiding publication bias entirely, even under well-intentioned circumstances.

Analogously, the evaluation of prescription medicines is also susceptible to a similar bias due to a preset statistical criterion (e.g., a false positive error rate below $\alpha = 2.5\%$).

To explore the magnitude of such bias, we can calculate this effect under certain assumptions. For instance, assume the effects of test drugs entering clinical trials follow a normal distribution. This allows us to quantify how regression to the mean might influence treatment effect estimates, improving our understanding of its impact on clinical outcomes.

Given the normal test statistic $Z = \dfrac{X - \mu}{\sigma\sqrt{n/2}}$ is the standard normal distribution with large sample size n, the conditional expectation on the statistical significance at a level of α is

$$E\left(Z\middle|Z\right\rangle z_{1-\alpha}\right) = \frac{\phi\left(z_{1-\alpha}\right)}{1 - \Phi\left(z_{1-\alpha}\right)}.$$

Thus, the expectation of the standardized bias, Bias $= \dfrac{X - \mu}{\sigma}$, can be obtained through linear transformation, and for $\alpha = 2.5\%$:

$$\text{Bias} = \sqrt{\frac{2}{n}}\,\frac{\phi(z_{1-\alpha})}{\alpha} = \sqrt{\frac{2}{n}}\,\frac{0.05844}{0.025} = \frac{3.306}{\sqrt{n}}.$$

This bias can be viewed as publication bias due to the requirement of statistical significance.

For typical sample sizes in clinical trials—ranging from 10 per group to 100 and up to 2500 per group—the upward bias in treatment effect estimates varies accordingly: approximately 1.05, 0.33, and 0.067 times the standard deviation of the responses, or around 2.34 times the standard error. In comparison, a typical clinical trial's standardized effect size falls within 0.8 (requiring $n=30$ per group for 85% power), 0.43 ($n=100$ per group for 85% power), and 0.1 ($n=1800$ per group for 85% power). In other words, bias of the published medical effect due to the statistical significance requirement approximately ranges from 67% to 130%.

In practical terms, this suggests that the mean treatment effects reported on drug labels in the U.S. are, on average, about twice the true effect size under given assumption. While we are aware of this systematic bias due to the statistical significance requirement, identifying which specific drugs overestimate or underestimate their treatment effect remains challenging.

7.5.2 Potential Bias Introduced by Funding

The relationship between research funding and potential bias is a critical and nuanced issue, particularly in areas like climate change research. Funding incentives often create positive feedback loops that reinforce particular narratives, potentially introducing biases into research and policy. This reinforcement can lead to publication bias and flawed policy decisions, as funding disproportionately supports certain perspectives while sidelining others.

In climate change research, for example, significant federal funding has been allocated over the years. According to a 2018 report by the U.S. Government Accountability Office (GAO, 2018), the Office of Management and Budget (OMB) reported over $154 billion in federal funding for climate change activities since 1993, distributed across various government agencies. This widespread investment has raised questions about the potential for fragmentation, overlap, and duplication in research efforts. Notably, there is no evidence of government funding explicitly supporting skepticism or exploring the bias introduced by government funding itself.

Funding inherently influences research agendas and priorities, though this influence does not necessarily equate to bias. However, it raises concerns about objectivity and the independence of science. A particularly significant issue is the reinforcement of consensus, wherein funding structures inadvertently favor prevailing scientific paradigms. Agencies are more likely to allocate resources to research aligned with the majority view, which can crowd out alternative perspectives.

The reinforcement of consensus can be quantitatively studied through analogy to the pick-the-winner design as discussed in Chapter 3.

7.5.2.1 Examples of Potential Bias

Critics of current funding practices highlight areas where bias may arise:

- In climate science, the majority of funding aligns with the consensus on human-induced climate change, with little room for skeptical perspectives. A 2005 study by McKitrick and Michaels suggested that government funding in climate research correlates with findings supporting the consensus, though this remains contentious.
- During the COVID-19 pandemic, substantial funding focused on vaccine development and efficacy. Critics argued that studies questioning vaccine risks or the effectiveness of mandates faced greater difficulty obtaining funding or publication opportunities.
- In pharmaceutical research, government and private funding has been scrutinized for favoring studies showing positive drug efficacy results, potentially underreporting negative outcomes. This demonstrates how funding incentives can shape research priorities and outcomes.

7.5.2.2 Research Outcomes on Funding Bias

Studies and reviews have explored the subtle effects of funding on research outcomes:

- Meta-analyses: Ioannidis (2005) found that financial conflicts of interest, including government funding, can subtly influence research outcomes, though not necessarily in an overt or intentional manner.
- Systematic reviews: Research shows that studies funded by specific industries, such as fossil fuels or pharmaceuticals, are more likely to produce favorable results. While government funding aims for neutrality, similar patterns could emerge if funding priorities influence research goals.

In conclusion, understanding and addressing potential bias introduced by government funding is essential for preserving the integrity and objectivity of science. While government funding has been pivotal in advancing research, its influence on agenda-setting and outcomes warrants careful scrutiny. Promoting transparency, balanced funding, and openness to diverse perspectives can help mitigate bias and enhance trust in publicly funded research. This ensures that science remains a robust, impartial tool for addressing society's most pressing challenges.

7.6 Friendship Paradox and Sampling Bias

The friendship paradox is a fascinating phenomenon in social network theory that reveals an unexpected statistical insight: On average, most people have fewer friends than their friends do. The paradox reveals inherent biases in how we perceive our social circles based on network structures.

It occurs because of sampling bias: Individuals with more friends are more likely to be sampled when looking at a person's friends. This means that friends with higher social connections disproportionately influence the average number of friends.

In a network, the degree of a node represents the number of connections (friends) it has. If the individual average is calculated by the average number of friends per person in the entire network and friends' average is the average number of friends among a person's friends, the paradox states that typically, the friends' average is higher because nodes with more connections are sampled more frequently.

The paradox can be proved mathematically as follows:

Assume friendships are mutual. Let k be the degree of a person, and $P(k)$ be the probability that a randomly selected person has k friends. The average degree is

$$\underline{k} = \sum_k kP(k).$$

The average number of friends among your friends is

$$\underline{k}_{\text{friends}} = \frac{\underline{k^2}}{\underline{k}}.$$

Since $\underline{k^2} \geq \underline{k}^2$ (by Cauchy–Schwarz inequality), it follows that

$$\underline{k}_{\text{friends}} \geq \underline{k}.$$

We use a four-friend network as an example to illustrate the paradox in Figure 7.3, where Alice, Bob, Cyrus, and Dai have 2, 2, 3, and 1 friend, respectively. The average number of friends in the network is 2, and the average number of friends of friends is 2.25.

In epidemiology, the paradox implies that identifying highly connected individuals can help in controlling disease outbreaks more effectively. In marketing, it means targeting influential individuals with large social networks can enhance the spread of information or products. In terms of social perception, this paradox can contribute to feelings of inadequacy, as individuals may perceive themselves as having fewer social connections compared to their peers.

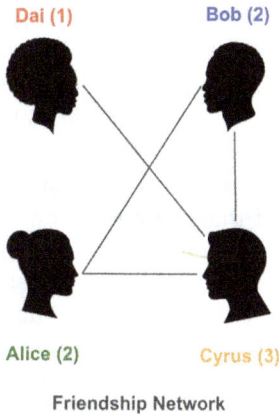

Number of Friends

Total number friends (connections):

   2(Alice) + 2(Bob) + 3(Cyrus) + 1(Dai) = 8

Average number of friends = 8/4 = **2**

Number of Friends of Friends

Alice: Bob(2) + Cyrus(3) = 5

Bob: Alice(2) + Cyrus(3) = 5

Cyrus: Alice(2) + Bob(2) + Dai(1) = 5

Dai: Cyrus(3) = 3

Total number of friends of friends = 18

Average number of friends of friends = 18/8 = **2.25**

Friendship Network

FIGURE 7.3
The friendship paradox: why my friends have more friends.

Through analogy, the paradox can be extended beyond the number of friends to other attributes like wealth or influence, where individuals tend to have peers with higher values in those attributes.

To address the sampling bias that leads to the friendship paradox, researchers and practitioners can use weighted sampling by applying weights based on node degrees that can help in obtaining more representative samples that account for the inherent bias toward highly connected nodes.

7.7 Berkson's Paradox: Restrictive Sampling Bias

Berkson's paradox or Berkson's bias (Berkson, 1946) is a type of selection bias that occurs when there is a spurious (false) association between two independent variables due to how data is sampled or selected. The paradox arises when the sample used to analyze the relationship between two variables is biased in such a way that it leads to the illusion of a negative correlation between variables that are actually independent.

Here are how the paradox playout:

1. Independent variables in the population: In the general population, the two variables X (diabetes) and Y (hypertension) are independent. The probability of having diabetes does not affect the probability of having hypertension and vice versa.

2. Selection condition: Suppose individuals are selected into the sample (hospitalized) based on the condition that they have either X or Y (or both). This forms a restrictive sampling process. In other words, the sample only includes people who have either diabetes or hypertension or both, but does not include people who have neither.

3. Induced negative correlation: In the selected sample, having one condition (say, diabetes) lowers the likelihood that the person has the other condition (hypertension) because they were already likely to be admitted due to the first condition. This creates a false negative correlation between X and Y in the selected sample, even though X and Y are independent in the general population.

Simply put, when you focus only on hospitalized patients, the sampling procedure effectively "filters out" individuals who do not have diabetes and hypertension. Therefore, within this restricted group, it appears as though people with diabetes are less likely to have hypertension (and vice versa), creating a false negative correlation.

Statistically, let us assume two events A and B represent two conditions (e.g., diabetes and hypertension), and C represents a condition where an individual is selected into a sample based on having either A or B or both. Berkson's paradox shows how conditioning on C (selection bias) can introduce a spurious negative correlation between A and B. Here is a mathematical explanation:

Define a selection condition C such that a person is selected if they have either condition A or condition B, or both: $C = A \cup B$.

$$P(A \cap B \mid C) = \frac{P(A \cap B \cap C)}{P(C)} = \frac{P(A \cap B)}{P(C)} = \frac{P(A \cap B)}{P(A) + P(B) - P(A \cap B)}.$$

Because $(A) + P(B) - P(A \cap B) \leq 1$, $P(A \cap B \mid C) \geq P(A \cap B)$, which suggests that, in the sample, the events A and B appear to be more mutually exclusive than they are in the general population. In other words, the restrictive sampling process will create a spurious negative correlation between A and B in the sample.

Berkson's paradox is a powerful illustration of how selection bias can distort perceived relationships between variables. Whether the paradox induces a spurious positive or spurious negative correlation depends on the nature of the selection criteria.

Berkson's paradox can be applied via analogies as shown below:

Education and employment: Suppose you study the relationship between the education level and employment status in a sample that only includes individuals who are employed. In the general population, education and employment might be positively correlated (more education increases the chance of employment). However, in a sample that includes only employed individuals, you might see a negative correlation. This is because, among employed people, those with lower education levels might have been admitted to the sample (employed) due to compensatory factors like exceptional skills, leading to the false conclusion that lower education correlates with higher employment.

Dating preferences: Imagine you are studying the relationship between attractiveness and wealth in individuals who are dating. In the general population, attractiveness and wealth might be independent or even positively correlated. However, among people who are currently dating (most likely either attractive or wealthy), you might observe a negative correlation because people who are less attractive may compensate by having more wealth, and vice versa. The selection bias comes from only observing people who are in the dating pool, not the general population.

University admissions: Suppose you study the correlation between two academic abilities (e.g., math and verbal skills) among students admitted to a top university. In the general population, these skills might be independent or positively correlated. However, because students with strong math skills but weak verbal skills (or vice versa) may still get admitted due to compensatory strengths in the other area, you might observe a negative correlation between math and verbal skills in the admitted students.

Wealth and risk: If we make analogies, in economics, we might study the relationship between wealth (condition *A*) and risk tolerance (condition *B*) among a group of individuals who have made large financial investments (selection criterion *C*) and conclude that among this sample, it might appear that individuals with higher wealth are less likely to take financial risks, while those with lower wealth are more likely to engage in riskier investments. This negative correlation is artificial, as it arises from focusing only on people who are selected for having made large investments. In the general population, wealth and risk tolerance could be uncorrelated or even positively correlated.

Artificial intelligence and bias in data selection: If we make analogies, in machine learning, Berkson's paradox can emerge when models are trained on datasets that are inherently biased or pre-filtered based on specific criteria. For example, imagine an AI system that is trained to predict job performance (condition *A*) based on educational background (condition *B*), but the dataset only includes employees who were hired based on having either a good educational background or prior work performance. Berkson's paradox in AI: The model may learn that job performance and education are negatively correlated because of how the data was selected (biased sample), even though there may be no such relationship in the general population. This spurious correlation could lead the model to incorrectly devalue educational qualifications in future predictions.

In summary, Berkson's paradox is a statistical pitfall that arises from selection bias and can lead to the false appearance of a negative correlation between independent variables. It underscores the importance of understanding the sampling process in any analysis, as conclusions drawn from biased samples may be entirely misleading. This paradox is particularly relevant in fields like epidemiology, economics, and social sciences, where data is often collected based on specific conditions (e.g., hospital patients, employed individuals, etc.). Recognizing and addressing this bias is crucial to making accurate inferences. Beyond Berkson's paradox, any restricted sampling procedure can introduce bias and the generalization of the conclusions based on the restricted sample to a general population can be misleading.

7.8 Cultural Impacts of Overinterpretation

Overinterpretation is a phenomenon where individuals or societies assign meanings or significance to works, ideas, or events that exceed their creators' original intentions. While often viewed critically, overinterpretation has profound cultural implications, influencing collective identity, historical narratives, and the trajectory of art, science, and politics. Here, we explore its cultural impacts further.

1. **Shaping national identity:** Overinterpretation often plays a central role in constructing and reinforcing national identity. For example, during the Cultural Revolution in China (1966–1976), Mao Zedong's poetry was elevated far beyond its literary quality, becoming a tool for political and cultural unification. By attributing grand ideological meanings to Mao's words, the regime instilled a sense of collective purpose and loyalty to his vision. This reshaping of artistic and intellectual content into cultural propaganda illustrates how overinterpretation can serve as a mechanism for social cohesion and political control.

2. **Elevating ordinary works to canonical status:** Through overinterpretation, ordinary works can be recontextualized and incorporated into the cultural canon. For example, historical figures, texts, or artifacts often gain heightened significance over time as they are viewed through contemporary lenses. This process can enrich cultural heritage, turning otherwise mundane works into symbols of collective memory or inspiration. The reinterpretation of Shakespeare's plays in modern contexts, for instance, has kept his works relevant and resonant across centuries, solidifying his status as a literary icon.

3. **Generating new forms of cultural expression:** Overinterpretation fuels creativity by inspiring derivative works, reinterpretations, and new forms of cultural expression. For example:

   - In art, overinterpretation of abstract works often leads to the creation of new art movements or critical theories.
   - In literature, speculative interpretations of an author's intent can inspire entirely new genres or adaptations.
     This interpretive process allows cultures to continuously reinvent themselves, maintaining a dynamic and evolving cultural landscape.

4. **Legitimizing ideologies and power structures:** Overinterpretation can influence ideologies or power structures by embedding them within the cultural fabric. For example, religious texts frequently undergo layers of interpretation that adapt their teachings to align with the prevailing socio-political context, reflecting and sometimes reinforcing the role of religious institutions.

5. **Obscuring original intent:** While overinterpretation can enrich cultural narratives, it can also obscure the original intent of creators. This distortion may lead to the loss of historical accuracy or the misrepresentation of a creator's vision. For example, overinterpreting historical texts to fit modern ideologies can strip them of their context, reducing their value as historical documents.

6. **Perpetuating cultural myths:** Overinterpretation often creates or perpetuates cultural myths, which can both unify and mislead societies. For instance, the over-idealization of figures like George Washington or Mahatma Gandhi has turned them into near-mythical symbols, simplifying their complexities to fit nationalistic narratives. While these myths inspire, they also risk creating unrealistic standards or obscuring nuanced truths.

7. **Stimulating intellectual inquiry:** On a positive note, overinterpretation can stimulate intellectual inquiry by encouraging debate and critical thinking. When a work is subjected to multiple interpretations, it often sparks discussions about its meaning, context, and relevance. This process can deepen societal understanding and promote intellectual engagement with cultural artifacts.

8. **Influencing global perception:** Overinterpretation can also shape how cultures are perceived globally. When specific aspects of a culture are magnified and reinterpreted, they often become symbolic of the culture as a whole. For example, the overemphasis on samurai culture in Japan or Confucian philosophy in China shapes global perceptions, sometimes at the expense of understanding these societies' broader diversity.

To summarize, overinterpretation, while often viewed as distorting or excessive, plays a crucial role in shaping cultural narratives and identities. It transforms the way societies understand their history, art, and ideologies, often leading to enriched cultural expression and global influence. However, it also carries risks, such as distorting original intentions, perpetuating myths, or reinforcing power structures. A nuanced understanding of overinterpretation reveals its dual role as both a creative force and a potential source of cultural misrepresentation.

7.9 Summary: Recognizing and Addressing Bias through Analogies

This chapter delves into the pervasive influence of biases in science and society, offering insights into their identification and mitigation through critical thinking. By exploring phenomena like confirmation bias, regression to the mean, survivorship bias, and funding bias, this chapter highlights how biases skew research outcomes and decision-making. It emphasizes strategies such as transparent methodologies, robust experimental designs, and the critical evaluation of sampling processes to neutralize individual and systemic biases. Clinical trials are recognized as a high-standard experimental method that can reduce various biases, as illustrated in Figure 7.4.

Key Takeaways

1. Bias recognition: The chapter emphasizes the importance of identifying hidden biases that influence observation, research, and

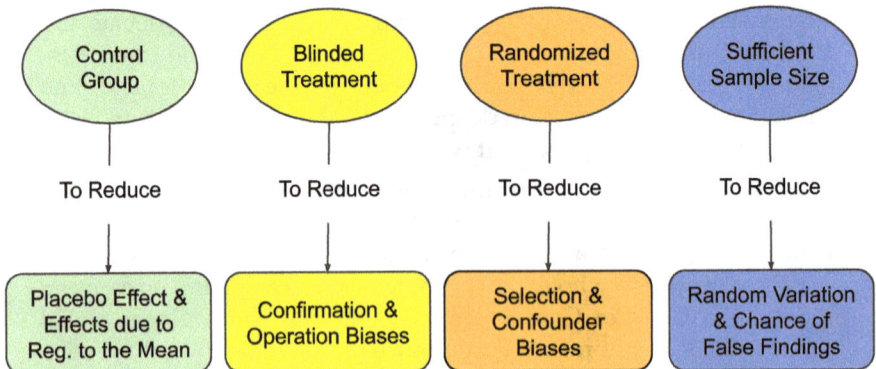

FIGURE 7.4
Clinical trial method can reduce various biases.

conclusions, such as the systemic effects of publication bias and funding incentives.

2. Confirmation bias: It is illustrated through cherry-picking data in research, political opinions, and dietary beliefs, providing relatable examples of how biases shape understanding.

3. Analyzing systemic issues: By critically examining examples like Berkson's paradox, friendship paradox, and survivorship bias, the chapter illustrates how flawed sampling and restrictive selection criteria can distort results.

4. Promoting objectivity: The chapter advocates for practices like open science, replication of studies, and balanced funding to counteract biases and enhance the reliability of research findings.

By integrating critical thinking and creative analogies, this chapter provides a nuanced exploration of bias, offering practical strategies to recognize, evaluate, and mitigate its effects. This chapter emphasizes the importance of collective effort and transparency in achieving unbiased, reliable research and decision-making.

Exercise

1. Provide an example of confirmation bias in scientific research. How could implementing double-blind studies or randomization minimize its effects?

2. Explain the phenomenon of regression to the mean using an example from sports or academics. How can this bias affect decision-making?

3. What critical thinking ideas stand out to you in this chapter?

4. Discuss the role of transparency and replicability in reducing biases in research. Propose a protocol for minimizing publication bias in a scientific study.

5. Explore how funding allocation can introduce bias in research priorities. Discuss whether creating a balanced funding system would reduce this bias or create new challenges.

6. Using the friendship paradox, explain how sampling bias might distort perceptions in social media. How could weighted sampling improve the representativeness of data?

7. Create a real-world analogy for Berkson's paradox in a non-medical field (e.g., hiring practices or marketing). Explain how this bias could mislead decision-makers.

8. Debate whether prioritizing research with positive results over null findings is justifiable. What are the potential impacts on scientific progress and public trust?

9. Discuss whether overinterpretation is more beneficial or harmful to society. Consider examples from literature, art, and historical narratives.

10. Analyze a historical event or industry trend that demonstrates survivorship bias. What lessons can be learned to avoid similar misinterpretations in the future?

11. Propose a research methodology or system that integrates the identification and mitigation of multiple biases (e.g., confirmation bias, funding bias, and sampling bias).

8

The Interesting yet Challenging Nature of Statistical Modeling

Analogies are fundamental to human understanding, offering profound insights into complex phenomena while fostering interdisciplinary connections and innovation.

8.1 Resampling Methods: Recursive Analogies

Resampling is a broader term encompassing any method that involves repeatedly sampling data points to assess the accuracy of sample statistics, validate models, test hypotheses, or perform sensitivity analysis. Resampling includes techniques such as bootstrap, cross-validation, permutation tests, and jackknife. Resampling methods are used to assess the accuracy of sample statistics, validate models, test hypotheses, or perform sensitivity analysis.

Bootstrap draws samples with replacement, cross-validation splits the data into multiple training and testing sets to validate model performance, permutation tests shuffle data labels to test the null hypothesis of no association, and jackknife systematically leaves out one observation at a time to estimate the statistic's variability.

The bootstrap method is simple, yet a powerful statistical tool that leverages resampling to estimate the distribution of a sample statistic, like a mean or variance, and it is particularly helpful in cases where traditional assumptions (e.g., normality and large sample size) might not hold.

The bootstrap method, introduced by Brad Efron (1979), involves repeatedly sampling from the observed data with replacement to create multiple bootstrap samples. Each bootstrap sample is the same size as the original dataset, and for each sample, we compute the statistic of interest. Repeating this process many times generates an empirical distribution of the statistic. From this distribution, we can estimate confidence intervals, standard errors, and even hypothesis tests without assuming an underlying population distribution. This is particularly useful when working with small datasets or data that does not meet parametric assumptions.

The term bootstrap itself is a metaphor—referring to the idea of pulling oneself up by the bootstraps. This speaks to the method's reliance on the sample data alone, without needing external parameters. This is achieved through

DOI: 10.1201/9781003630081-8

self-similarity and iteration (recursive analogy): The bootstrap process uses small, iterative steps to gain insight from within, much like recursion.

Uncertainty and confidence (flexible exploration): Bootstrapping builds confidence intervals directly from the sample, without making strong distributional assumptions. Creatively, this reflects a flexible approach to exploring uncertainty and seeking solutions even when traditional or well-defined paths are not available. It is a reminder that, sometimes, being less rigid with assumptions (about data or ideas) can lead to more reliable insights.

The plug-in principle, which uses parameter or distribution replacement, is a typical application of analogy. The principle states that to estimate $\theta(F)$ when F is unknown, you can "plug in" the empirical distribution $\hat{F}_n$ (constructed from your sample) in place of F. This yields the estimate $\theta\left(\hat{F}_n\right)$ as a natural approximation of $\theta(F)$. From analogy perspective, the plug-in principle keeps the structure $\theta(\cdot)$, but replaces the parameter F by $\hat{F}_n$.

To learn more about why and how resampling with replacement can approximate real-life variation, how uncertainty can be measured even with limited data, and why bootstrapping does not rely on strict theoretical assumptions, making it versatile, you can read further on books by Efron and Tibshirani (1993), Efron and Hastie (2016), Chernick (2007), and Chang (2011b).

8.2 Frauds Detected by Benford's Law—Analogies in Action

In many practical scenarios, simple statistical principles can outperform complex theories, much like a flashlight revealing a hidden path in the dark. Benford's law exemplifies this by uncovering hidden anomalies through a straightforward yet effective approach. It is observed that in many naturally occurring datasets (e.g., electricity bills, stock prices, or populations), leading digits follow a predictable, non-uniform distribution or Benford's law.

Benford's law reflects the natural order in real-world numbers. Varian (1972) highlighted its potential for fraud detection, noting that fabricated data often assumes uniform digit distributions, diverging from natural patterns. Nigrini (1999) later expanded its use in detecting accounting and expense fraud, showing how violations of simple patterns reveal deeper irregularities.

While not a definitive proof of fraud, Benford's Law acts as a litmus test, flagging anomalies for further investigation. Its reliability has been demonstrated in U.S. courts and cases like the 2009 Iranian elections. In the 2009 Italian municipal elections, forensic analysts applied Benford's law to vote counts, finding deviations—such as underrepresentation of "1" and overrepresentation of higher digits—raising suspicions of electoral fraud and prompting further scrutiny (Roukema, 2014).

Benford's law states that in many naturally occurring datasets, the probability $P(d)$ that a number's leading digit is d, where $d \in \{1, 2, \dots, 9\}$ is given by:

$$P(d) = \log_{10}(1 + 1/d).$$

This means smaller digits (like 1) occur more frequently as the leading digit than larger ones (like 9).

For instance:

- The digit "1" appears as the leading digit approximately 30.1% of the time.
- The digit "9" appears much less frequently, around 4.6% of the time.

This counterintuitive pattern reflects a broader analogy to natural phenomena: randomness often conceals an inherent order (Figure 8.1). By comparing observed data to the expected distribution, Benford's law serves as a mirror, exposing discrepancies between fabricated and authentic data (Chang, 2012).

The core justification for Benford's Law lies in the principle of scale invariance (see Chapter 9). For the distribution of leading digits to be universal—untied to arbitrary units like kilometers or miles, dollars or euros—rescaling the data should not alter the distribution. Among scale-invariant functions, the logarithm is the simplest and most commonly applied.

In any numerical base b, the probability of the leading digit d (where $1 \le d < b$) is generalized as:

$$P(d) = \log_b(1 + 1/d).$$

This supports the claim that the law's form depends on logarithmic properties rather than specific numbers.

In conclusion, while Benford's Law was initially discovered in specific fields, its applicability extends across diverse domains through the use of analogies. Whether analyzing financial data, election results, or socio-economic metrics,

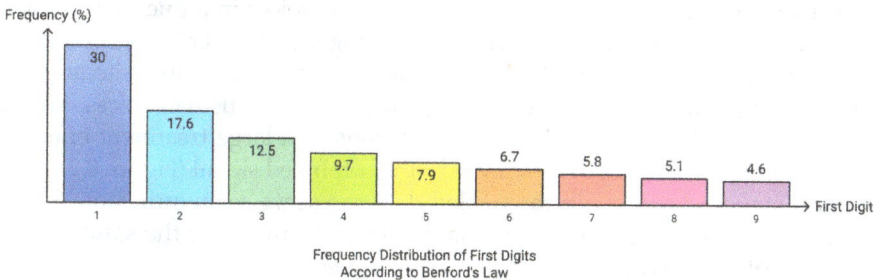

Frequency Distribution of First Digits
According to Benford's Law

FIGURE 8.1

Frequency distribution of digits following Benford's law.

Benford's Law exemplifies how fundamental principles, when guided by analogy, can reveal hidden patterns and provide unexpected insights into complex problems.

8.3 Drug Effect Size Revealed from Blinded Data: An Analogy of Mixed Signals

Gaussian mixture models (GMMs), which are built on the concept of a mixture of normal distributions, provide a powerful analogy for deciphering hidden structures within complex data. Much like how a blend of distinct musical notes creates a composite chord, GMMs represent data as a combination of multiple normal distributions, each with its own mean and variance. This probabilistic model is particularly insightful when data originates from multiple underlying subpopulations, where each subpopulation can be modeled independently by its own distribution. By capturing the interplay between these distributions, GMMs enable us to uncover subtle patterns and relationships that would remain hidden with simpler statistical methods.

This analogy becomes especially pertinent in clinical trials. It is often assumed that the treatment difference between a test drug and a placebo cannot be discerned from pooled (blinded) data. However, this belief overlooks the potential for indirect inference. When pooled data are modeled as a mixture of two normal distributions with differing means, the overall distribution can reveal underlying truths. If the difference between the means is significant, the pooled distribution becomes bimodal, signaling two distinct subpopulations. Conversely, when the means are close, the signal fades into a unimodal distribution.

8.3.1 Structural Analogy: Real-Life Cases of Mixed Normals

The real-world application of this concept is akin to detecting blended but distinct patterns, such as the combination of two voices in a duet. Each voice maintains its own unique pitch and tone, yet together they create a composite harmony. Similarly, in clinical trials, the pooled data act as the composite, and the underlying treatment and placebo groups represent the distinct voices. When these "voices" diverge significantly—analogous to a large treatment effect—their separation becomes evident even in the combined signal (Figure 8.2).

For instance, skewness in the pooled data can serve as an indicator of this divergence. Skewness, calculated using a formula involving the sample size fractions of the two components

$$\Gamma = w_1 w_2 \left(w_2 - w_1 \right) \delta^3 ,$$

Mixtures of Two Normal Distributions with Different Treatment Effects

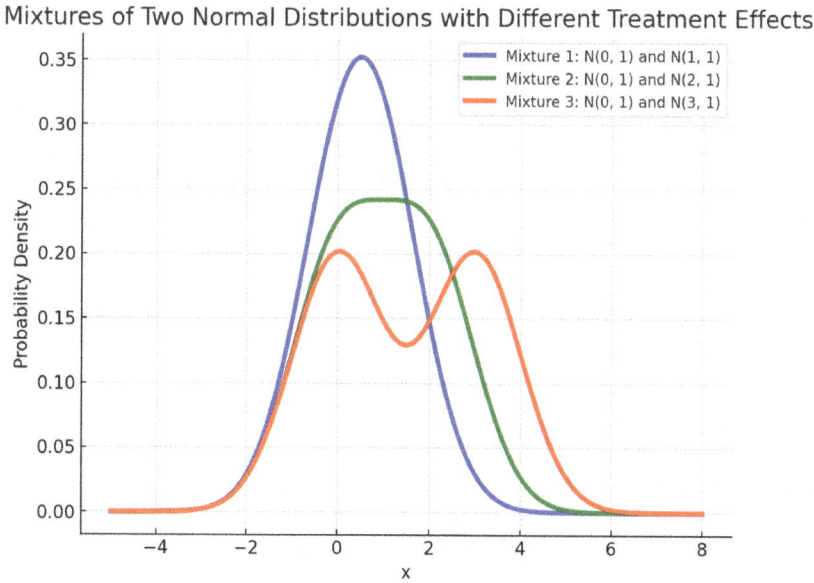

FIGURE 8.2
Normal mixture distributions: visualizing overlapping data.

acts like a tuner for detecting the extent of separation between the distributions. In a randomization ratio of 1:3, the skewness formula simplifies to reveal the treatment difference with the formula:

$$\hat{\delta} = 4\left(\frac{\Gamma}{6}\right)^{1/3},$$

using the plug-in principle (Chang, 2019). Other sophisticated techniques, such as the expectation-maximization (EM) algorithm, provide further tools for analyzing these mixtures without explicitly unblinding the data.

8.3.2 Beyond Clinical Trials: Mixed Normals in Action

The analogy of mixture distributions extends far beyond clinical trials. In technology, healthcare, finance, and environmental science, mixture normal models are used to reveal hidden structures in complex datasets. For example:

- In finance, they can model asset returns from different market regimes.
- In healthcare, they identify patient subpopulations with differing responses to treatment.
- In environmental science, they detect distinct climate patterns within aggregated weather data.

In conclusion, the use of GMMs provides an elegant analogy for understanding and analyzing data originating from multiple subpopulations. By interpreting bimodal or skewed distributions as signals of underlying differences, researchers can infer key insights from pooled data without explicit separation. This analogy not only applies to clinical trials but extends to a wide range of real-world phenomena, reflecting the power of statistical models to uncover hidden truths in blended datasets.

8.4 Controversies of Statistical Models

8.4.1 A Simple yet Multifaceted Model

Statistical modeling is like trying to understand the mechanics of a clock without taking it apart. Every gear (or variable) interacts in intricate ways, and the model you choose determines how you perceive those interactions. Suppose we aim to predict weight using height. In the first attempt, we consider total height (H_1) alone. To refine the model, we divide height into upper body height (H_2) and lower body height (H_3), producing the following formulation:

$$\text{Model 1: Weight} = a_2 H_2 + a_3 H_3 + \text{RE}.$$

Here a_2 and a_3 are coefficients representing the contributions of H_2 and H_3, and RE is the residual error. Then, recognizing that $H_2 = H_1 - H_3$, we reformulate the model:

$$\text{Model 2: Weight} = a_2(H_1 - H_3) + a_3 H_3 + \text{RE}.$$

Rearranging further, we derive:

$$\text{Model 3: Weight} = a_2 H_1 + (a_3 - a_2)H_3 + \text{RE}.$$

While these three models are mathematically equivalent and fit the same data, they lead to different interpretations of the lower body effect:

- In Model 1, it is interpreted as a_3.
- In Model 3, it is interpreted as $a_3 - a_2$.

This subtle difference in formulation can drastically shift conclusions, even leading to opposite signs for the lower body effect. This analogy demonstrates how a model's structure can shape the narrative, despite identical mathematical accuracy (Figure 8.3).

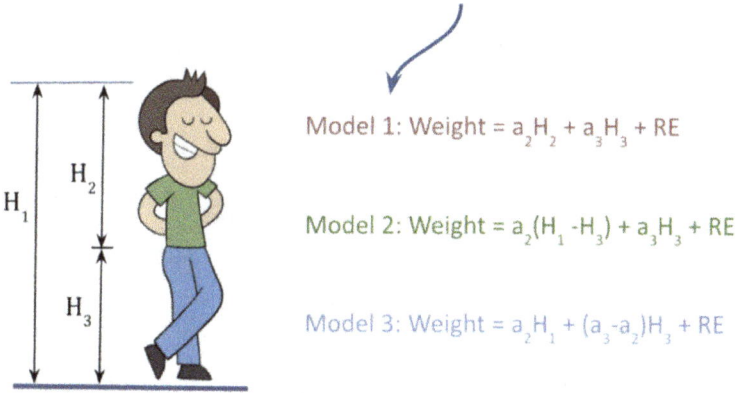

Three mathematically equivalent models fitting the same data with the same random error.

Model 1: Weight = $a_2 H_2 + a_3 H_3$ + RE

Model 2: Weight = $a_2(H_1 - H_3) + a_3 H_3$ + RE

Model 3: Weight = $a_2 H_1 + (a_3 - a_2)H_3$ + RE

FIGURE 8.3
Attribute effect controversies in data interpretation.

8.4.1.1 Analogy: Clinical Trials and the Puzzle of Variable Relationships

Now, imagine that weight represents a patient's response to a medical treatment in a clinical trial, and H_1 and H_3 represent correlated genomic and phenotypic attributes, with H_3 being the treatment variable. Even when H_3 is randomized (as treatments are in clinical trials), H_1 might remain unbalanced due to the impossibility of perfectly balancing infinite factors under randomization.

This imbalance can create an observed association between H_1 and H_3, muddying the analysis. Like the height models above, different formulations (e.g., Model 1 and Model 3) could lead to conflicting conclusions about the treatment effect—one suggesting benefit, the other harm.

8.4.1.2 Balancing Statistical Rigor and Real-World Relevance

When independent variables like genes or biomarkers are involved, alongside treatments, the model can mathematically isolate the treatment effect, stating that it is assessed "keeping H_1 or H_2 constant." However, this approach raises a practical question: If you were a patient, would you take the drug or not, and based on which model? This tension between mathematical precision and practical relevance underscores the limitations of statistical modeling as a decision-making tool.

8.4.1.3 Orthogonality and the Limits of Interpretability

If H_1 and H_3 are orthogonal (statistically independent), there will be no association between them, simplifying the analysis. In such cases, tools like principal component analysis (PCA) can replace H_1 with a statistically independent

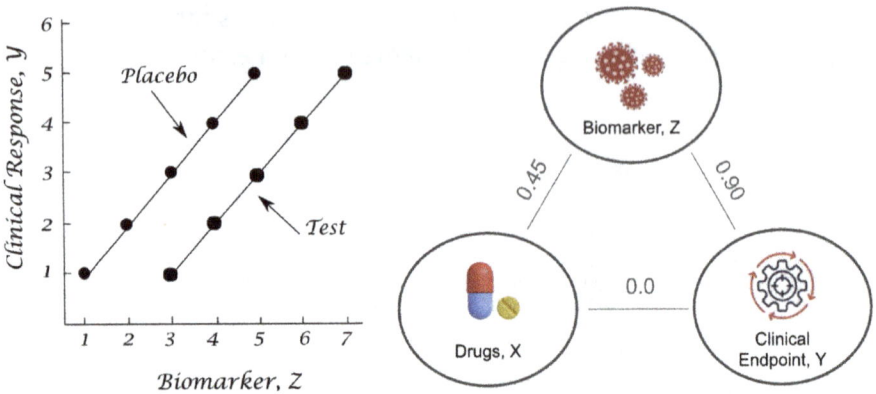

FIGURE 8.4
Intransitivity of correlations.

component. However, this new component might lack clear interpretability, akin to replacing a clock gear with an abstract shape that fits perfectly but leaves you wondering how it functions.

8.4.2 The Non-Transitivity of Correlation

The relationships among variables are often counterintuitive. For example, as shown in Figure 8.4,

- Suppose treatment (X) has a moderate correlation (0.45) with a bio-marker (Z).
- The biomarker (Z) strongly predicts the clinical response (Y) with a correlation of 0.9.
- However, there is no direct correlation between the treatment (X) and the clinical response (Y).

This lack of transitivity highlights the complexity of indirect relationships and the challenges of interpretation. The treatment might influence the biomarker, which in turn predicts the clinical response, but the direct link between treatment and response remains hidden. This situation is analogous to interpreting friendships: a friend of your friends is not necessarily your friend.

8.5 Strategy of Optimal Choices

8.5.1 Description of the Secretary Problem

The secretary problem, also known as the marriage problem, was discussed informally in earlier decades (Lindley, 1961).

Imagine a dating game where a female participant seeks to choose the best possible partner from a pool of 100 male candidates ($n = 100$). The candidates are presented to her one at a time in random order. After meeting each candidate, she must decide immediately whether to accept him as her partner or reject him and move on to the next candidate. Importantly, she cannot return to a previously rejected candidate, and she has only one chance to select.

The goal is to maximize the probability of selecting the absolute best candidate from a pool of 100 options. Making optimal choices in sequential decision-making scenarios can be particularly challenging, as options are presented one at a time, and decisions must be made without knowledge of future possibilities. However, a simple yet mathematically optimal strategy exists: observe and reject the first 37% of options and then select the next candidate who surpasses all those previously evaluated.

This approach, known as the strategy of optimal choice and often exemplified by the 37% rule, provides a robust framework for maximizing the probability of selecting the best option. Remarkably, this strategy yields a consistent success probability of $1/e$ (approximately 37%), regardless of the size of the candidate pool. This counterintuitive result demonstrates the power of mathematics in solving complex real-world decision-making problems.

The power of mathematics lies in its ability to address and simplify complex real-world decision-making challenges. The **37% rule** is not merely a theoretical concept; it has practical applications in a wide array of real-life scenarios where optimal stopping decisions are crucial.

8.5.2 Probabilistic Model of the Optimal Strategy

Objective: Maximize the probability of selecting the best candidate (option) out of n sequentially presented candidates.

Constraints:

- Sequential presentation: Candidates are interviewed one at a time in random order.
- Immediate decision: After each interview, an immediate and irreversible decision must be made to accept or reject the candidate.

Let r be the number of candidates to observe and reject before considering selection. Our goal is to determine the value of r that maximizes the probability $P(r)$ of selecting the best candidate.

Let E_k be the event that the best candidate appears at position k and S_k be the candidate selected as the best candidate at position k. Since the candidates are presented in random order, the probability of E_k is:

$$P(E_k) = 1/n.$$

Therefore,

$$P(r) = \frac{1}{n} \sum_{k=r+1}^{n} P\left(S_k \mid E_k\right).$$

Given that the best candidate is at position k, the probability that we select them is equal to

$P\left(\text{observed maximum among the first } k-1 \text{ candidates is within the first } r\right)$.

This implies:

$$P(S_k \mid E_k) = \frac{r}{k-1}.$$

$$P(r) = \frac{1}{n} \sum_{k=r+1}^{n} \frac{r}{k-1} = \frac{r}{n} \sum_{k=r+1}^{n} \frac{1}{k-1} = \frac{r}{n} \sum_{k=r}^{n-1} \frac{1}{k}.$$

When n is large, we can approximate the sum using the integral of the harmonic series.

$$H_m = \sum_{k=1}^{m} \frac{1}{k} \approx \ln(m) + \gamma, \text{ where Euler–Mascheroni constant approximately} = 0.5772.$$

Thus,

$$P(r) = \frac{r}{n}\left(H_n - H_r\right) \approx \frac{r}{n} \ln \frac{n}{r}.$$

Taking the derivative of $P(r)$ with respect to r, and letting it to zero, it leads to

$$\frac{1}{n}\left(\ln\left(\frac{n}{r} - 1\right)\right) = 0.$$

Solving for r: $r = n/e$.

Using $r = n/e$ in $P(r)$ expression to obtain the maximum of P:

$$P = \frac{n/e}{n} \ln \frac{n}{n/e} = 1/e \approx 36.8\%.$$

This result holds true regardless of the total number of candidates n, provided n is sufficiently large.

8.6 Summary: Analogies and Statistical Modeling as Pathways to Insight

Statistical models, much like analogies, simplify complex phenomena to enhance understanding. However, the way a model is structured influences the story it tells, sometimes leading to different interpretations despite mathematical equivalence. This chapter explores how analogies have driven statistical innovation—from the bootstrap method's recursive logic to Benford's law's unexpected fraud detection applications. It also highlights the interpretive challenges posed by statistical models, such as the ambiguity of mathematically equivalent formulations and the non-transitivity of correlation. Through real-world examples—including optimal stopping strategies, GMMs, and resampling techniques—this chapter underscores the power of analogical thinking in revealing hidden structures, guiding decision-making, and balancing statistical precision with practical insight.

Key Takeaways

1. Interpreting statistical models with precision: Mathematically equivalent models can lead to different interpretations, emphasizing the importance of contextual understanding in statistical analysis.

2. Resampling and recursive analogies: The bootstrap method and other resampling techniques provide robust ways to estimate uncertainty without strict parametric assumptions.

3. Benford's law as an analytical tool: Benford's law detects anomalies in numerical datasets, aiding fraud detection and forensic analysis.

4. GMMs and hidden structures: GMMs can uncover treatment effects from blinded data in clinical trials, finance, and environmental science.

5. The 37% rule and optimal stopping strategies: The secretary problem illustrates how the 37% rule optimizes sequential decision-making in hiring, investments, and other real-world scenarios.

6. Non-transitivity of correlation: Correlations do not always transfer across variables, complicating causal inference and statistical interpretation.

Exercise

1. What critical thinking ideas and elements of creative analogies stand out to you in this chapter?

2. Describe how the bootstrap method uses the analogy of "pulling oneself up by one's bootstraps." Propose a scenario where resampling would be advantageous over traditional parametric methods.

3. Apply Benford's law to a hypothetical dataset (e.g., financial records or election data). Identify patterns that may indicate fraud or anomalies and explain the statistical reasoning behind your conclusions.

4. Develop a creative analogy to explain regression or classification in machine learning to someone without a technical background. Ensure your analogy is both accessible and accurate.

5. Use the analogy of a duet to explain GMMs and how they identify subpopulations in data. Provide a real-world example where GMMs could reveal hidden structures.

6. Debate whether the structure of a statistical model influences its interpretation more than the data itself. Use examples from the chapter, such as different formulations of a weight prediction model.

7. Discuss whether analogies oversimplify complex phenomena or enhance understanding. Provide examples from the chapter to support your argument.

8. Using the secretary problem and the 37% rule, describe a real-world situation where this strategy could be applied. Discuss the situations where such a rule does not apply.

9. Explore the concept of non-transitive relationships in a dataset. Create a hypothetical example where variable A influences B, and B influences C, but A and C have no direct correlation. Explain how this could affect interpretation.

9

Dimensional Analysis-Assisted
Statistical Modeling

The scale invariance principle asserts that physical laws should be independent of the units used to measure physical quantities.

Dimensional analysis leverages analogies to create unit-consistent, scalable statistical models.

9.1 Introduction to Dimensional Analysis

Dimensional analysis is fundamentally based on the invariance principle—the idea that physical laws should be independent of the units used to measure physical quantities. This is a form of scale invariance, ensuring that the mathematical expressions describing physical phenomena remain valid regardless of the choice of units.

Dimensional analysis is a powerful tool for simplifying and scaling complex problems, enabling efficient modeling, experimentation, and insight generation without requiring full-scale replication. It allows engineers and scientists to address challenges such as designing a spaceship to the moon or studying fluid behavior in scaled-down water dam models. By relying on dimensionless parameters, such as Reynolds number, engineers can model spacecraft performance in wind tunnels or simulations without replicating the moon's gravity or atmospheric conditions. Similarly, small-scale models of dams provide critical insights into structural behavior and fluid dynamics without constructing full-sized systems.

This approach naturally provides substantial cost savings by allowing small-scale or computational experiments to replace expensive full-scale testing. It enables predictive modeling of real-world behavior in extreme or inaccessible environments, such as planetary surfaces or massive engineering projects.

The Buckingham pi theorem further streamlines analysis by reducing complex systems into independent dimensionless parameters, making experimentation and calculations more manageable.

DOI: 10.1201/9781003630081-9

Dimensional analysis can also serve as a powerful analogy for statistical modeling by helping identify key variables, relationships, and interactions in data. Here is how dimensional analysis can be applied as an analogy in statistical modeling.

9.1.1 Identifying Core Variables and Relationships

- In dimensional analysis, a complex physical phenomenon is broken down into fundamental units (e.g., mass, length, and time). Similarly, in statistical modeling, you can identify core variables that drive relationships in the data.
- For instance, in modeling a biological system, key variables might include age, dose, and response. Like fundamental units, these variables can guide model formulation by focusing on the most influential factors.

9.1.2 Creating Dimensionless Ratios

- Dimensional analysis often involves forming dimensionless ratios that reveal invariant properties across scales. In statistical modeling, creating standardized or normalized ratios helps compare variables across different scales and units, highlighting essential relationships.
- For example, forming ratios of variables (such as risk ratios or odds ratios in clinical trials) can reveal proportional effects, helping compare treatments or conditions regardless of the original scale.

9.1.3 Reducing Complexity

- By reducing equations to dimensionless forms, dimensional analysis can simplify the number of variables in a system. In statistics, this is akin to dimension reduction techniques like principal component analysis (PCA), which identify combinations of variables (like principal components) that explain the most variance in the data.
- This helps create simpler models by focusing on a reduced set of composite variables or ratios rather than individual measurements.

9.1.4 Scaling Relationships for Model Extrapolation

- Dimensional analysis allows predictions across scales (e.g., from a lab to real-world applications) by understanding scaling laws. Similarly, in statistical modeling, once relationships are established, they can generalize to new data or extrapolate to different conditions by understanding how core variables scale.

- For instance, in dose-response models, scaling doses appropriately could allow predictions at untested dose levels.

9.1.5 Checking Model Consistency and Units

- Dimensional analysis verifies that units are consistent across equations. In statistical modeling, ensuring unit consistency (for example, matching scales of predictors and responses) can prevent errors in interpretation and improve model reliability.

Dimensional analysis is a powerful tool in engineering modeling, but rarely used in statistical modeling. Dimensional analysis helps simplify complex problems, check the consistency of equations, and derive scaling laws. When viewed from an analogy perspective, dimensional analysis can be seen as an analogy-making process where units (dimensions) of physical quantities are treated similarly to how we make connections between ideas in different contexts.

9.2 Dimensional Analysis: A Framework for Drawing Analogies

9.2.1 Mapping Dimensions to Concepts

In dimensional analysis, physical quantities like speed, force, and energy are expressed in terms of fundamental dimensions (e.g., length $[L]$, mass $[M]$, and time $[T]$). These fundamental dimensions can be seen as the building blocks, just as concepts in different fields of science can be reduced to foundational ideas.

The process of dimensional analysis involves mapping these complex quantities back to their simplest components, much like how analogies map new, complex ideas to more familiar concepts.

For example, the quantity force is expressed as $F = ma$ (mass times acceleration), which in terms of dimensions is $[F] = [M][L][T]^{-2}$. Similarly, when we create an analogy, we are reducing the complexity of a new concept (force) by linking it to more familiar ideas (mass and acceleration).

9.2.2 Dimensional Homogeneity as a Consistency Check

Dimensional analysis ensures dimensional homogeneity, meaning that the units on both sides of an equation must match. This principle serves as a consistency check in equations, much like how analogies provide a way to check for logical consistency across different domains.

If the dimensions on both sides of an equation do not align, it signals that something is wrong with the formulation—similar to how a faulty analogy (like "sun is to moon as water is to fire") would signal a mismatch in logic or structure.

9.2.3 Scaling Laws and Similarity

Dimensional analysis is heavily used to derive scaling laws, which describe how one physical quantity changes relative to another (e.g., how the force on a car changes as its size changes). In this sense, it is a form of analogy-making, where similarities are used to predict outcomes under different conditions.

Analogies in science often involve finding scaling similarities between phenomena (e.g., how the behavior of gases can be analogous to the behavior of traffic flow). In analogy, scaling might involve saying, "Just as a small increase in pressure can lead to turbulent flow in a liquid, a small increase in stress can lead to chaos in a social system," drawing parallels between different domains.

9.2.4 Dimensional Analysis and Cross-Disciplinary Analogies

Analogies often bridge gaps between different disciplines. Dimensional analysis does something similar by providing tools that work across various physical systems, regardless of the specific field (e.g., physics, biology, or engineering). An analogy could be: "Fluid dynamics is to aerodynamics as electromagnetism is to circuits," where the underlying principles of flow and behavior in one field inform the understanding of another.

9.3 Scaling Laws in Engineering and BioMed Engineering

9.3.1 Modeling and Simulations in Engineering

In wind tunnel testing for aerodynamics, before building an airplane, engineers often test scale models in a wind tunnel. The Reynolds number, a dimensionless quantity derived through dimensional analysis, helps ensure that airflow in the model resembles the airflow around the full-sized plane. By ensuring that the Reynolds number in the model matches that of the actual airplane, engineers can predict the aerodynamic performance of the plane at full scale.

In earthquake simulations, dimensional analysis is used to scale down seismic forces and vibrations so that small-scale models of buildings or infrastructure can be tested in laboratories. The similarity principles derived from dimensional analysis ensure that the scaled-down model behaves in the same way as the full-scale structure would during an actual earthquake, helping engineers improve designs for seismic safety.

9.3.2 Most Comfortable Speed to Walk

It is evident that the most comfortable walking speed of a human depends on their size. Generally, adults walk faster than children when walking alone. This raises an interesting question: how does size affect the walking speed of geometrically similar individuals? While personal preference can influence walking speed, there is a specific speed at which the energy expenditure per unit distance is minimized—this is the most comfortable walking speed. To explore this relationship, we can apply dimensional analysis.

Let v represent the most comfortable walking speed, measured in meters per second (m/s), and L the characteristic linear size (e.g., height) in meters (m). We also consider gravitational acceleration, g, measured in meters per second squared (m/s^2), as it influences the vertical movement of the body's center of mass during walking and thus affects energy expenditure. Including g also generalizes the results to different celestial bodies, such as the Moon or Earth.

With three variables (v, L, and g) and two fundamental dimensions (length and time), we form one dimensionless variable:

$$\Pi = \frac{v}{\sqrt{Lg}}.$$

Since Π must be a constant, we derive:

$$v = c\sqrt{Lg},$$

where c is a dimensionless constant. This implies that the most comfortable walking speed is proportional to the square root of a person's height or characteristic length.

Further assuming geometric similarity and, as a first approximation, that human density is uniform, the body mass M is proportional to L^3. Substituting this relationship, we find:

$$v \propto M^{1/6}.$$

Thus, the most comfortable walking speed is proportional to the sixth root of an individual's mass (Figure 9.1).

9.3.3 Allometric Scaling of the Heart Rate

Variable dimensions:

- Heart rate (f): Number of heartbeats per unit time with dimension: $[f] = [T^{-1}]$
- Body mass (M) with dimension: $[M] = [M]$
- Metabolic rate (B): Energy expended per unit time with dimension: $[B] = [M \cdot L^2 \cdot T^{-3}]$
- Stroke volume (V_s): volume of blood pumped per heartbeat with dimension: $[V_s] = [L^3]$

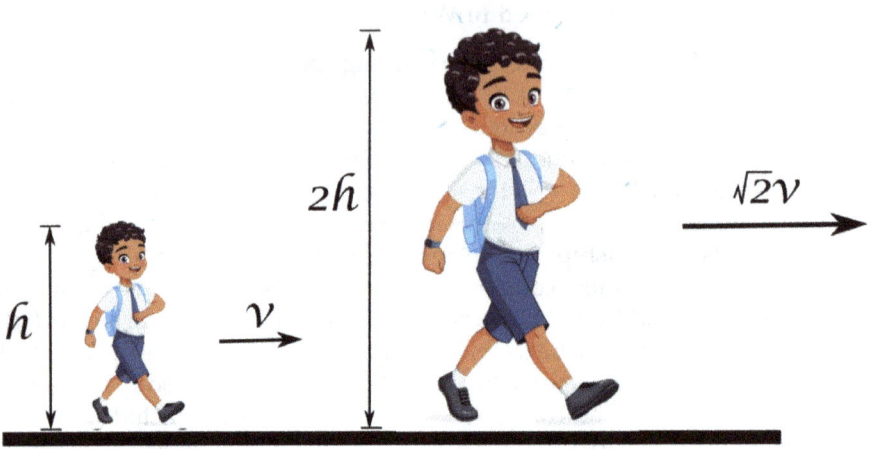

Most comfortable speed to walk.

Assumptions:

1. Metabolic rate scaling: $B \propto M^{3/4}$
2. Geometric similarity: $L \propto M^{1/3}$
3. Stroke value scaling: $V_s \propto M$
4. Energy per heartbeat proportional to the stroke volume: $E_h \propto V_s \propto M$
5. Metabolic rate relation: $B = E_h \cdot f$

Derivation:
$B = E_h \cdot f \propto M \cdot f$, that is, $M^{3/4} \propto M \cdot f$. Solve for heart rate: $f \propto M^{-1/4}$.
Equivalently, the heart rate can be expressed as

$$f = cM^{-1/4},$$

where c can be obtained through statistical modeling using experimental or observational data.

The heart rate scales inversely with the quarter power of body mass: $f \propto M^{-1/4}$. This indicates that larger animals have slower heart rates compared to smaller animals, aligning with empirical observations in biology.

9.3.4 Metabolic Rate and Body Size (Kleiber's Law)

In biology, metabolic rate R of animals is known to scale with body mass M. The relationship between metabolic rate and body mass, crucial for understanding energy use in animals, can be derived from dimensional analysis.

Metabolic rate R has dimension $[M][L^2][T^{-3}]$. Body mass has dimension $[M]$. If we assume that an animal's surface area (which is proportional to $M^{2/3}$) is related to heat loss and metabolic rate, we can propose that: $R \propto M^{2/3}$.

This scaling law is derived based on the assumption that heat loss (and thus energy consumption) is proportional to the surface area of the animal. However, the experimental data show a slightly different relationship: $R \propto M^{3/4}$. This is known as Kleiber's law, which we have used in the derivation of allometric scaling of the heart rate.

9.4 Buckingham Pi Theorem for Simplifying Modeling

The Buckingham pi theorem is a key theorem in dimensional analysis that allows us to reduce the number of independent variables in a physical problem by combining them into dimensionless groups called pi terms (π terms). This reduction simplifies the analysis of complex systems and helps in the formulation of generalized equations that can be applied universally.

Buckingham π theorem: If you have a physical problem described by a relationship involving n variables and k fundamental dimensions (e.g., mass $[M]$, length $[L]$, and time $[T]$), then the variables can be grouped into $(n - k)$ dimensionless parameters (π terms).

The theory can be used to:

- Simplification: Reduces the number of variables, simplifying experiments and modeling.
- Generalization: Dimensionless groups are universal, allowing results to be applied to different systems.

9.4.1 Steps to Apply the Buckingham Pi Theorem

1. List all relevant variables: Identify all the independent and dependent variables influencing the system.
2. Express variables in fundamental dimensions: Write the dimensions of each variable using fundamental units (e.g., $[M]$, $[L]$, and $[T]$).
3. Determine the number of fundamental dimensions (k): Count the number of fundamental dimensions involved.
4. Calculate the number of dimensionless groups (π terms): Number of π terms $= n - k$.
5. Choose repeating variables:
   - Select k variables that include all fundamental dimensions.
   - Repeating variables should not form a dimensionless group among themselves.

6. Form the dimensionless groups:
   - Combine the repeating variables with each of the remaining variables to form π terms.
   - Ensure that each π term is dimensionless.
7. Write the final dimensionless equation: Express the relationship between the π terms.

Let us discuss the simplest example of applying the Buckingham pi theorem in biology: modeling drug diffusion through a membrane.

The goal is to determine how the drug diffusion Flux (J): $\left[ML^{-2}T^{-1}\right]$ through a membrane depends on:

- Diffusion coefficient (D): How quickly the drug diffuse $\left[L^2T^{-1}\right]$
- Concentration difference (ΔC): Difference in drug concentration across the membrane $\left[ML^{-3}\right]$
- Membrane thickness (δ): Distance the drug travels $[L]$

Total variable (n)=4
Fundamental dimensions (k)=3 (M, L, T)
Number of dimensionless groups=$n-k=1$

9.4.2 Application of the Pi Theorem

1. Select repeating variable $(D, \Delta C, \delta)$ covering all dimensions.
2. From dimensionless group (π) with the remaining variable (J):

$$\pi = \frac{J \cdot \delta}{D \cdot \Delta C}.$$

The dimensionless group simplifies the relationship to:

$$\frac{J \cdot \delta}{D \cdot \Delta C} = c, \quad \text{or equivalently,} \quad J = \frac{D \cdot \Delta C}{\delta} \times c,$$

This matches Fick's first law of diffusion when the constant $c = -1$, while in general, c can be obtained through statistical modeling using experimental or observational data.

By using the Buckingham pi theorem, we have reduced four variables into a single dimensionless group, simplified the complex relationship governing drug diffusion, and demonstrated how fundamental principles can streamline modeling in biology and medicine. This approach aids in understanding and predicting how drugs permeate biological membranes, which is essential for drug development and therapeutic effectiveness.

9.5 Summary: Dimensional Analysis and Its Role in Statistical Modeling

This chapter explores the application of dimensional analysis as both a practical tool and a conceptual analogy for enhancing statistical modeling. Borrowed from physics and engineering, dimensional analysis simplifies complex systems by identifying fundamental units, creating dimensionless ratios, and uncovering scaling relationships. The chapter illustrates how these principles apply to diverse fields, such as modeling drug diffusion, determining optimal walking speeds, and understanding biological scaling laws like Kleiber's law.

Key Takeaways

1. Simplification of complexity: Dimensional analysis reduces complex problems into manageable components by creating dimensionless groups, enabling deeper understanding and streamlined modeling in physics, engineering, and biology.

2. Buckingham pi theorem: This method's reduction of variables into dimensionless groups serves as a metaphor for identifying universal patterns across disciplines.

3. Analogies for dimensionless ratios: Tools like Reynolds numbers in fluid dynamics or drug diffusion models illustrate how dimensionless parameters reveal universal properties across systems.

4. Scaling as analogy: Scaling principles, such as predicting walking speeds or heart rates based on body mass, demonstrate parallels between physical laws and biological phenomena.

Exercises

1. Provide an example of a real-world equation where dimensional consistency is essential. Explain how inconsistencies in dimensions could lead to errors in modeling or predictions.

2. Explain how dimensional analysis is used to determine the most comfortable walking speed for a human. How does the relationship between walking speed and body size reflect fundamental principles?

3. What critical thinking ideas and elements of creative analogies stand out to you in this chapter?

4. Discuss the use of scaling laws in engineering or biology. Provide an example where scaling principles are applied to extrapolate experimental findings to real-world scenarios.

5. Using the Buckingham pi theorem, derive a dimensionless relationship for fluid flow through a pipe. Include variables such as flow velocity, pipe diameter, fluid viscosity, and density.

6. Create an analogy to explain the concept of dimensionless groups (e.g., Reynolds number) to someone unfamiliar with the concept. Ensure the analogy simplifies the idea while maintaining its scientific integrity.

7. Draw a parallel between dimensional analysis in engineering and a non-scientific field (e.g., economics or sociology). How can this analogy aid in understanding complex systems?

8. Debate whether dimensional analysis can be effectively applied to fields outside physical sciences, such as social sciences or behavioral studies. Support your argument with examples.

9. Discuss the limitations of dimensional analysis as a modeling tool. How does it compare to other modeling techniques like regression or machine learning?

10. Derive the dimensionless group for drug diffusion through a membrane using the Buckingham pi theorem. How does this result align with Fick's first law of diffusion?

11. Using Kleiber's law $(R = M^{3/4})$, explain why larger animals have lower metabolic rates per unit of body mass compared to smaller animals. Extend this idea to explore how this scaling principle might influence drug dosage in medical treatments.

10

Recursive Functions as Self-Referencing Analogies

Self-referencing analogies in recursive functions often lead to paradoxes, but these apparent contradictions can provide profound clarity about the underlying structures.

10.1 Self-Referential Paradox as Hierarchical Analogy

Many things we are using in reasoning are analogies, but we may not realize it. For example, recursive functions, whether structural, numerical, or conceptual, operate as hierarchical analogies. Given $f(x)$, we draw an analogy between $f(x)$ and x itself, which naturally leads to the next level, $f(f(x))$. This process continues with $f(f(f(x)))$, and so on, establishing an iterative analogy within the function's output and its input.

In recursion, $f(x)$ can represent diverse structures: logical statements, geometric or data configurations, or even scientific principles. When recursion involves a statement, it creates self-referential expressions, as seen in classic paradoxes like the liar paradox ("This statement is false") and Russell's barber paradox. When recursion applies to geometric structures, it forms geometric patterns like fractals, producing self-similar shapes (Figure 10.1).

Let us look into various self-referencing paradoxes:

The liar paradox: "This statement is false." If the statement is true, then it must be false as it claims, but if it is false, then it must be true. This creates a paradox because the statement contradicts itself, leading to a logical inconsistency.

The paradox of probability: No theory is absolutely correct, meaning the fundamental probability theory (in the frequentist or Bayesian paradigm) itself cannot have a certainty of correctness (its probability is less than 1). This raises an intriguing paradox: if we acknowledge the uncertainty in the foundational theory of probability, should we adjust every probability we calculate to account for this meta-uncertainty?

DOI: 10.1201/9781003630081-10

This illustration of **minimalist cartoon-style** features clean lines, muted colors, and a simplified yet expressive depiction of the subject. This particular style emphasizes simplicity while conveying the concept effectively, as seen in the recursive "self-reference" theme.

FIGURE 10.1
The self-reference paradox explained.

The barber paradox: In a town, there is a barber who shaves everyone who does not shave themselves. Does the barber shave himself? If the barber shaves himself, then according to the rule, he should not be shaving himself. If he does not shave himself, then by the rule, he must shave himself. This leads to a paradox—a simpler version of Russell's paradox.

Russell's paradox: Consider the set of all sets that do not contain themselves. Does this set contain itself? If the set contains itself, then by definition it should not contain itself. If it does not contain itself, then by definition it should contain itself. This creates a paradox within naive set theory. Russell's paradox led to the development of more rigorous set theories, such as the Zermelo–Fraenkel set theory, to avoid such contradictions.

Paradox of promise: Our ethical standard dictates that we should keep our promises. But if someone never promised to keep their promises in the first place, would it then be acceptable for them to break their promises?

The paradox of democracy is the inherent tension between freedom of choice and the need to preserve the democratic system. On the one hand, democracy relies on the principle that individuals should have the freedom to make choices, including the choice to reject democracy itself. On the other hand, allowing such a choice could lead to the erosion of democracy, resulting in authoritarianism or other non-democratic systems.

This creates a dilemma:

- If democracy permits the choice of non-democracy, it risks self-destruction.
- If democracy prohibits such a choice, it undermines its own foundational principle of freedom.

The paradox forces democracies to balance unrestricted freedom with the need to protect the system's core values, raising questions about whether a democracy can remain democratic while restricting certain choices to ensure its survival.

A country might practice democracy at a national level while enforcing dictatorship internationally, or maintain a domestic dictatorship while advocating for democracy abroad. However, if a dictator initiates reforms aimed at transitioning toward democracy, does this act itself reflect the exercise of dictatorship or an embrace of democratic principles? This raises a complex question about the nature of power and intent: is the reform an extension of authoritarian control or a genuine step toward empowering the people (Chang, 2014)?

Discriminant analysis is a widely used approach in statistics, serving as a tool to distinguish one group with specific attributes from others. More broadly, statistical modeling and scientific inquiry essentially aim to identify and differentiate patterns among groups. However, as Chang (2014) emphasizes, the pursuit of scientific truth must not overshadow ethical considerations. The following example employs self-referencing to construct a paradox that highlights the ethical complexities inherent in such analyses.

The discrimination paradox draws an analogy to the self-referential paradox, applying its recursive logic to arguments about discrimination—creating a paradoxical and seemingly unresolvable cycle. Consider statements tied to a specific attribute, such as age group, race, or gender, like: "The drug is more effective for women under X years old," "People of attribute X commit more crimes than others" or "People with attribute X have been unfairly discriminated against." Such statements inherently involve stereotyping or generalization. Consequently, any decision or policy based on these statements—regardless of its truth—constitutes an attribute-based decision. These decisions either discriminate against individuals with attribute X or favor them, potentially at the expense of others. Thus, policies that aim to combat discrimination through attribute-based generalizations risk falling into the same ethical dilemma posed by the trolley problem (Chapter 1): can a "noble intention" justify morally complex means?

What makes this paradox particularly intriguing is that similarity grouping—the very act of categorizing entities based on shared attributes—is foundational to science and the formulation of scientific laws. Without attribute-based generalizations, science as we know it would not exist. The use of concepts like "dog" or "chair" is itself a form of stereotyping: when we refer to "a chair," we generalize across countless variations of chairs based on shared features. Even on a personal level, recognizing the absence of a "type" of person in a group photo involves a form of stereotyping—it presupposes an attribute-based classification of people. Without such "stereotyping," AI will not be able to draw required images when she is asked to draw, e.g., an old man or an Asian baby.

This tension between the necessity of generalization in science and its ethical implications in societal contexts raises profound questions about how we define fairness, discrimination, and objectivity. The idea of similarity grouping will be explored further in Chapters 15 and 16, where we examine the role of attribute-based generalizations in scientific reasoning and their broader philosophical implications.

10.2 The Dilemma of Understanding

In national TV shows and social media, we often encounter arguments such as: The poor say, "You are rich—how could you possibly understand the concerns of the poor?" This statement implies that only someone who is poor can truly understand the struggles of poverty. Such remarks are often effective in silencing the opposing party.

But what if the rich were to respond with the same logic? "You're not wealthy—how can you know that the rich don't understand?" Each side could continue this recursive exchange indefinitely, creating a paradox that questions the validity of their arguments.

This debate raises a timeless question: Can a rich person genuinely understand the experiences of the poor, or are their perceptions merely projections of their own biases and interpretations? Similarly, can the poor truly understand the experiences and challenges of the wealthy?

The exchange highlights the reciprocal nature of understanding, illustrating that the very reasoning used to question another's perspective can just as easily be turned back on oneself. Ultimately, this challenges us to reflect on the nature of empathy and the limits of understanding across differing experiences and social divides.

Furthermore, stating that "rich people cannot understand poor people" or vice versa is a form of stereotyping, which serves as a foundation for discrimination—a topic we explore in greater depth in a later section.

This **scientific digital illustration** with a futuristic and conceptual aesthetic. It uses clean, precise lines, intricate networks, and a minimalist color palette to convey complexity and interconnectedness. The design incorporates neural and geometric patterns, blending anatomical accuracy with abstract, symbolic representations of cognitive processes and self-awareness. This style is highly detailed and emphasizes clarity and a modern, technological feel.

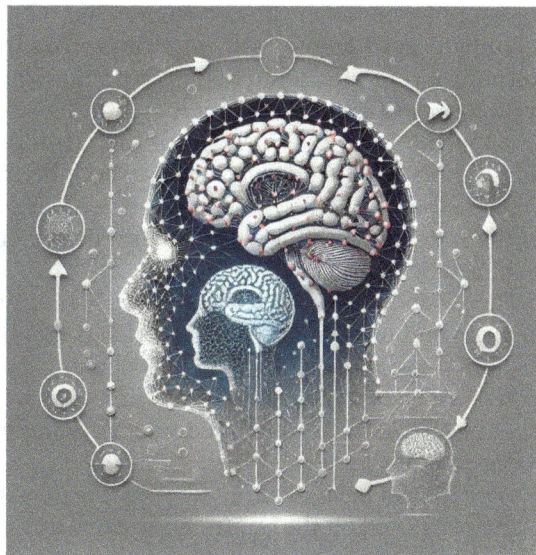

FIGURE 10.2
Recursive self-inclusive mind enabling awareness.

10.3 Self-Awareness as a Recursive Function and Hierarchical Analogy

In the context of an individual's mind, recursion enables self-awareness—a person can conceive of themselves within their own thoughts, illustrated as "a man within a man" (Figure 10.2). This mechanism is fundamental to humanized AI, where recursive self-reference supports an artificial form of self-awareness.

This recursive framework invites exploration of logical principles such as the pigeonhole principle, the principle of noncontradiction, and reductio ad absurdum to understand how recursion challenges and expands the boundaries of logical consistency and self-reference.

As an example, we are going to discuss proof by contradiction and how to use self-referencing to disprove it, that is, use proof by contradiction to invalidate general applicability of proof by contradiction.

10.4 Critique and Refutation of "Proof by Contradiction"

We first recall two principles: proof by contradiction and the pigeonhole principle.

Proof by contradiction, also known as reductio ad absurdum, is a logical technique used to prove a proposition true by assuming its negation and deriving a contradiction. This contradiction implies that the negation is false, thus confirming the proposition's truth. This method relies on the principle of noncontradiction, which states that a proposition (e.g., "A is true") and its negation ("not A is true") cannot both be true simultaneously. In simple terms, either "A is true" or "not A is true," but not both. For instance, only one of the statements "the stock market will rise" or "the stock market will not rise" can be true at a specific time.

The pigeonhole principle (Figure 10.3) is a fundamental concept in combinatorics and mathematics. It states that if you have more items than containers (or "pigeonholes") and you distribute those items among the containers, at least one container must contain more than one item. In formal terms, if $N+1$ or more objects are placed into N containers, then at least one container will contain at least two objects.

We can provide a disproof of reductio ad absurdum using the pigeonhole principle, based on the brain's finite capacity.

FIGURE 10.3
Illustration of the pigeonhole principle.

Just as Einstein used the speed of light as an ultimate limit to derive relativity, we apply the concept of limited brain states to derive similarly counter-intuitive results. Suppose the human brain has a finite number of states—say, 100 billion. A person cannot distinctly recognize more than 100 billion unique items because of the pigeonhole principle, which can be demonstrated by contradiction (note that memory also allocates states for emotions, perceptions of infinity, and more). This creates a loop: reductio ad absurdum proves the pigeonhole principle, which then discredits reductio ad absurdum itself.

The limits of brain states are different from computer memory. When a computer's memory is full, it can erase and replace content because we operate it with a larger memory capacity. However, brain-state limits mean that humans cannot differentiate an excess of distinct elements, making the world appear finite, even if one could live indefinitely. For instance, if a brain state associated with "sweet" is reached, one might experience sweetness even when tasting something salty.

Limited brain states constrain perception. If someone had only eight brain states, they could not reliably distinguish single-digit numbers and might perceive "8" and "9" as identical (Figure 10.4). Because both memory and time are finite, we cannot sustain self-referential thought (e.g., "thinking about thinking about thinking...") indefinitely. This self-referential limitation indicates that Aristotle's logic is a first-level system, not well-suited to recursive or self-referential thinking.

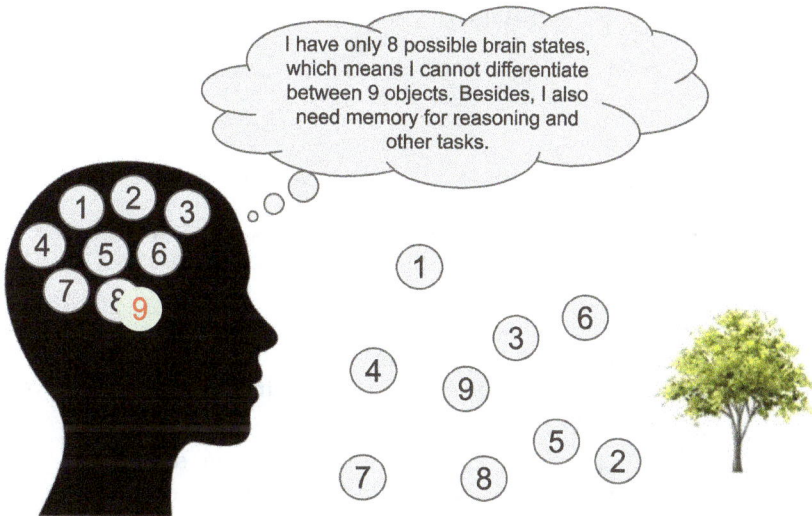

FIGURE 10.4
Perceptual limitations of brain states.

There are alternative ways to challenge that proof by contradiction is universally valid, such as based on the fact that statements can influence the outcome making two contradictory statements "true" in a nuanced way. Here is how this could play out:

1. Self-fulfilling prophecy: If a powerful figure says, "the stock market will rise," investor confidence may increase, causing the market to rise. Conversely, if they predict a decline, caution could prevent a rise. In either case, the chosen statement may influence reality.

2. Quantum logic analogy: Analogous to superposition in quantum mechanics, we might consider these statements in a "superposed" state, where both outcomes are possible until the influential statement "collapses" (or measured) one into reality.

3. Broader truth definition: Truth is that which cannot be disproven in theory, not merely what can be proven true.

4. Multifaceted world: Objective world is multifaceted, no unique truth

5. Social constructivism: In sociology, self-fulfilling prophecies allow for contradictory beliefs to coexist within separate groups, as discussed in Robert K. Merton's *The Self-Fulfilling Prophecy.*

Dynamic logic is a well-established concept in formal logic and computer science originated to model and reason about the behavior of systems that change over time, particularly in the context of computer programs and automated processes. Dynamic logic allows us to analyze actions, sequences of actions, and how they affect states or conditions.

10.5 Self-Similarity: Fractals

A fractal is a geometric shape or pattern that displays self-similarity and complexity at every scale. Some key characteristics of fractals are:

1. Self-similarity: Fractals appear similar at different scales, meaning that if you zoom in on a small part of the shape, it resembles the larger structure. This can be exact (perfect self-similarity) or approximate (statistical self-similarity).

2.. Infinite detail: Fractals contain infinitely intricate details. As you zoom in, new structures appear, and this process can continue indefinitely, theoretically.

3. Fractal dimension: Unlike traditional geometric shapes, fractals are measured with a fractal (or fractional) dimension, which can be a non-integer value. This dimension reflects the degree of complexity or roughness in the fractal structure. For instance, a line has a

dimension of 1, a square has a dimension of 2, but a fractal line can have a dimension between 1 and 2.

4. Iterative construction: Fractals are often generated through repeated, recursive processes. For example, in the famous Mandelbrot set, complex mathematical rules are applied repeatedly to create the fractal structure.

Examples of fractals with self-similarity include:

- Mandelbrot set: The Mandelbrot set is a mathematical fractal created by iterating a complex equation. Zooming in on the boundary reveals similar shapes at every scale, though the exact details vary slightly, showing self-similarity.
- Sierpinski triangle: This fractal is made by recursively removing triangles from a larger triangle. Each iteration creates smaller triangles within triangles, with the pattern repeating exactly at each scale (Figure 10.5).
- Koch snowflake: Starting with an equilateral triangle, the Koch snowflake is generated by adding smaller triangles to each side repeatedly. The shape never closes, creating infinite length along its perimeter while remaining bounded, and it shows exact self-similarity.

In addition to these fractals with mathematical self-similarities, there are fractals with statistical self-similarities, meaning geometrical shapes at different levels are similar in distribution. Brownian motion (Figure 10.5) is the most well-known fractal with self-similarities in distribution.

Sierpinski Triangle Fractals (Detho 5)

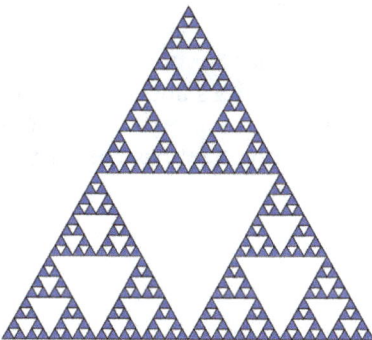

Simulated 2D Brownian Motion

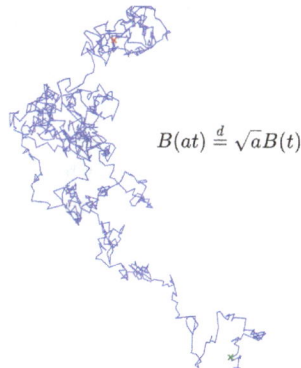

$$B(at) \overset{d}{=} \sqrt{a}B(t)$$

Mathematical Self-Similar Sierpinski Fractals

Statistical Self-Similar Brownian Motion

FIGURE 10.5
Mathematical vs statistical self-similarities.

10.5.1 Natural Fractals with Statistical Similarities

- **Coastlines**: The length and detail of coastlines appear similar regardless of the scale you view them at, making them a classic example of statistical self-similarity.
- **Clouds and mountains**: The roughness and shapes of clouds and mountains look similar up close or far away, exhibiting random self-similarity.
- **Trees and blood vessels**: Branching patterns in trees and blood vessels exhibit self-similarity, as smaller branches or vessels resemble the larger structure.

10.5.2 Fractals and Self-Similarity Applications

1. **Modeling natural phenomena**: Fractals model complex, natural structures, like clouds, mountains, and coastlines, which are self-similar and cannot be represented accurately by traditional geometry.
2. **Signal and image processing**: Fractal compression techniques leverage self-similarity, especially for images, by storing a small amount of data about self-similar patterns that can recreate complex images.
3. **Computer graphics**: Fractal algorithms generate realistic natural textures and landscapes in games and movies, using recursive algorithms to model the irregularities found in nature.
4. **Biology and medicine**: Self-similarity in fractals models branching structures, such as blood vessels, lung bronchi, and neurons, helping to analyze complex patterns in anatomy and physiology.
5. **Finance**: Fractal analysis is used to study market behavior, where price movements exhibit self-similar patterns over different time scales, aiding in identifying trends and volatility.

In addition to deterministic algorithms of similarity, in statistical self-similarity, the fractal does not repeat precisely but has a statistical similarity. Patterns or distributions resemble each other in a general sense, even though the specific details may differ. Brownian motion and coastlines are examples where the patterns are statistically similar at different scales.

10.6 Summary: Recursive Functions and Self-Referencing Analogies

This chapter explores recursive functions as self-referencing analogies, highlighting their role in logical reasoning, paradoxes, and practical applications. By delving into self-referential paradoxes the chapter examines how recursion both challenges traditional logical frameworks and provides a foundation for understanding self-awareness, complex systems, and fractals. The chapter underscores recursion's dual nature as both a source of paradoxes and a tool for modeling intricate structures.

Key Takeaways

1. Logic and Paradoxes: The chapter critiques conventional logical methods, such as proof by contradiction, using recursive reasoning to highlight their limitations. Paradoxes like Russell's paradox and the liar paradox illustrate the inherent tensions in self-referential systems, prompting re-evaluation of foundational assumptions.

2. Ethical dilemmas and social implications: Analogies like the discrimination paradox and the dilemma of understanding extend recursion to ethical and social contexts, challenging readers to assess how stereotyping, judgment, and systemic dynamics shape fairness and decision-making.

3. Modeling self-awareness: Recursive frameworks are essential for understanding self-awareness in both human cognition and artificial intelligence. Self-referential processes enable meta-thinking, or "thinking about thinking," and support the development of AI systems that mimic this complexity.

4. Mathematical and statistical self-similarity: Fractals like the Koch snowflake, Mandelbrot set, and Brownian motion demonstrate recursion, with mathematical fractals producing identical patterns and statistical fractals incorporating variability.

5. Self-referential analogies: Recursion is compared to a mirror reflecting itself, forming infinite loops that blend beauty with contradiction.

6. Applications of recursive systems: Fractal structures model natural and engineered systems, offering insights into diverse phenomena such as coastlines, blood vessel branching, and market behaviors. Recursive analogies bridge abstract concepts and practical applications.

This chapter emphasizes the interplay of critical thinking and creative analogies, illustrating how recursive functions challenge established logic while offering profound insights into the structure and behavior of complex systems.

Exercises

1. Explain the liar paradox and Russell's paradox in your own words. How do these paradoxes illustrate the challenges of self-referential systems?

2. Define fractals and list their key characteristics (e.g., self-similarity and infinite detail). Provide an example of a natural fractal and explain its significance.

3. What critical thinking ideas and elements of creative analogies stand out to you in this chapter?

4. Analyze the paradox of democracy: How does allowing the freedom to reject democracy challenge its foundational principles? Propose potential solutions to this paradox.

5. Discuss how recursion can enable self-awareness in artificial intelligence. Provide an example of a recursive process that might be used to simulate self-awareness.

6. Create your own analogy to explain recursion to someone unfamiliar with the concept. Use everyday objects or scenarios to illustrate self-referencing processes.

7. Identify a real-world application of fractals (e.g., computer graphics, medicine, or finance). Explain how their self-similarity aids in solving practical problems.

8. Debate whether logical systems like proof by contradiction should be reconsidered in light of self-referential paradoxes. Can alternative logical systems better address these challenges?

9. Discuss the ethical dilemmas of using attribute-based generalizations in science and society. Is it possible to avoid stereotyping while still making meaningful categorizations?

10. Using the pigeonhole principle as discussed in the chapter, critique the universal applicability of proof by contradiction. How might alternative logical frameworks address its limitations?

11. Explore the concept of dynamic logic. Provide an example of how it could be applied to a real-world problem involving systems that change over time.

11

Creative Data Visualization

Creative data visualization transforms data into compelling visual analogies, bridging abstract information and intuitive understanding.

11.1 Design Thinking in Data Visualization

Design thinking is a human-centered approach to problem-solving that emphasizes creativity, empathy, and iteration. In data visualization, it ensures that visualizations are functional, engaging, accessible, and tailored to the audience. This approach enhances clarity and impact, transforming raw data into meaningful insights.

In business and marketing, graphical design, a core element of user experience (UX) design, combines typography, color, imagery, and layout to communicate messages effectively across various mediums. These principles are equally applicable in scientific publications and presentations.

Data visualization focuses on representing data visually through charts, graphs, and maps. By applying graphical design principles, it balances aesthetics and functionality, ensuring data is both comprehensible and engaging for audiences in business and marketing contexts. In scientific communications, data visualization emphasizes a delicate balance between scientific rigor and simplicity, making complex data accessible without compromising accuracy.

11.1.1 Core Principles of Data Visualization

1. Clarity and simplicity: Effective visualizations prioritize concise and clear communication, avoiding unnecessary complexity or distractions.

2. Aesthetic appeal: Engaging designs capture attention and enhance memorability. Creativity can make even mundane data visually compelling.

DOI: 10.1201/9781003630081-11

3. Narrative and storytelling: The best visualizations guide viewers through a narrative, highlighting key insights and relationships within the data.

4. Cognitive considerations: Principles like Gestalt laws, color perception, and data density inform how viewers interpret visualizations, ensuring they are intuitive and not overwhelming.

11.1.2 Key Techniques for Effective Visualization

1. Choosing the right format: Match the data and its purpose with appropriate visual forms.

2. Balancing purpose and audience: Align visualizations with goals—whether to inform, persuade, or engage—while considering the needs and preferences of the target audience.

3. Blending art and science: Combine scientific accuracy with creative elements like stylized graphics or motion to deliver impactful and memorable visual narratives.

11.1.3 Drawing Attention in Data Visualization

Drawing attention in data visualization ensures key insights are effectively communicated. Effective attention-drawing balances emphasis with clarity, ensuring viewers focus on insights without being overwhelmed. The key techniques include:

- Color and contrast: Use distinct colors or contrasting shades to highlight critical data points or trends while muting less relevant elements.

- Annotations: Add callouts or labels to emphasize significant data points, anomalies, or trends directly within the chart.

- Hierarchy: Arrange elements to guide the viewer's focus, starting with prominent features (e.g., larger or brighter elements).

- Whitespace: Use negative space to isolate important data, enhancing its visibility.

- Interactivity: Allow user interactions like hovering, filtering, or drilling down to explore highlighted insights.

By integrating design thinking with graphical design principles and techniques, data visualization becomes a powerful tool for communication and analysis. This approach ensures that visualizations are not only visually appealing but also purposeful, intuitive, and capable of transforming data into meaningful stories.

11.2 Types of Data

To embrace scientific rigor, simplicity, and aesthetics in data visualization, a solid understanding of data and variable types is invaluable.

Variables or formatted data, whether raw or derived, can be categorized based on their type, format, and the kind of information they represent. Below is a structured overview of different types of variables:

1. **Data types:**
   - Binary: Two possible values (e.g., Yes/No).
   - Count: Non-negative integers (e.g., hospital visits).
   - Nominal: Categories without order (e.g., gender and blood type).
   - Ordinal: Ordered categories without consistent intervals (e.g., pain severity and socioeconomic class).
   - Interval: Numeric data with equal intervals but no true zero (e.g., temperature in Celsius).
   - Percentage: Normalized data to a base of 100 (e.g., vaccination rates).
   - Ratio: Relationships between two quantities (e.g., debt-to-income ratio).
   - Censored: Time-to-event data where some events have not occurred (e.g., patients alive at study end).

2. **Time-related aspects:**
   - Cross-sectional: Collected at a single time point (e.g., census results).
   - Longitudinal: Repeated data over time for the same subjects (e.g., health metrics).
   - Time-series: Sequential data indexed by time (e.g., daily stock prices).
   - Time-to-event (survival): Time until a specific event occurs (e.g., recovery after surgery).

3. **Statistical role:**
   - Independent variables: Variables manipulated to observe effects (e.g., drug dosage).
   - Dependent variables: Outcome affected by independent variables (e.g., recovery rates).
   - Control variables: Held constant to isolate effects (e.g., temperature in experiments).

4. **Aggregation:**
   - Aggregate data: Summarized group data (e.g., average income by region).
   - Individual-level data: Collected at the individual level (e.g., personal health records).

5. **Specialized types:**
   - Spatial: Represent geographic information (e.g., latitude/longitude).
   - Probabilistic: Represent probabilities (e.g., weather forecasts).
   - Multivariate: Multiple variables studied simultaneously (e.g., climate models).
   - Hierarchical: Nested relationships (e.g., students in classes in schools).

11.3 Data Visualization with Examples

Different types of graphs serve various data types and purposes. In terms of purpose, commonly used graph types can be categorized as follows:

1. Comparison: Bar chart, column chart, stacked bar/column chart, grouped bar chart
2. Distribution: Histogram, box plot, violin plot, density plot
3. Trend analysis: Line chart, area chart, step chart
4. Relationships: Scatter plot, bubble chart, heatmap
5. Composition: Pie chart, doughnut chart, tree map, stacked area chart
6. Ranking: Horizontal bar chart, lollipop chart, dot plot
7. Geospatial data: Choropleth map, point map, bubble map
8. Hierarchical data or clustering: Tree diagram, sunburst chart, clustered network graph
9. Multivariate data: Parallel coordinates plot, radar chart
10. Specialized/advanced: Gantt chart, candlestick chart, funnel chart

A circular barplot for a radial representation of data using polar coordinates, blending a bar chart with a circular layout for aesthetic appeal, a bubble chart for a scatter plot where the size and color of the points represent additional dimensions, adding richness to the visualization, a spiral plot for a creative visualization for cyclic or periodic data, representing data distribution along a spiral path for aesthetic and analytical appeal, and heatmap (Figure 11.1)

Heatmap Example

	X1	X2	X3	X4	X5	X6	X7	X8	X9	X10
Y1	0.24	0.32	0.5	0.24	0.1	0.87	0.54	0.79	0.13	0.7
Y2	0.44	0.69	0.59	0.62	0.86	0.3	0.8	0.51	0.89	0.1
Y3	0.49	0.075	0.51	0.2	0.52	0.95	0.41	0.23	0.77	0.76
Y4	0.94	0.2	0.71	0.094	0.91	0.68	0.56	0.6	0.1	0.89
Y5	0.13	0.35	0.62	0.59	0.87	0.92	0.63	0.47	0.77	0.032
Y6	0.19	0.54	0.4	0.42	0.63	0.18	0.23	0.31	0.9	0.85
Y7	0.14	0.0067	0.87	0.84	0.94	0.29	0.35	0.93	0.2	0.38
Y8	0.89	0.91	0.9	0.15	0.24	0.99	0.3	0.15	0.36	0.55
Y9	0.46	0.26	0.53	0.63	0.47	0.92	0.36	0.49	0.77	0.47
Y10	0.8	0.67	0.59	0.6	0.72	0.14	0.97	0.77	0.34	0.064

Safety Endpoints (y-axis) / Efficacy Endpoints (x-axis) / Intensity (color bar: 0.2, 0.4, 0.6, 0.8)

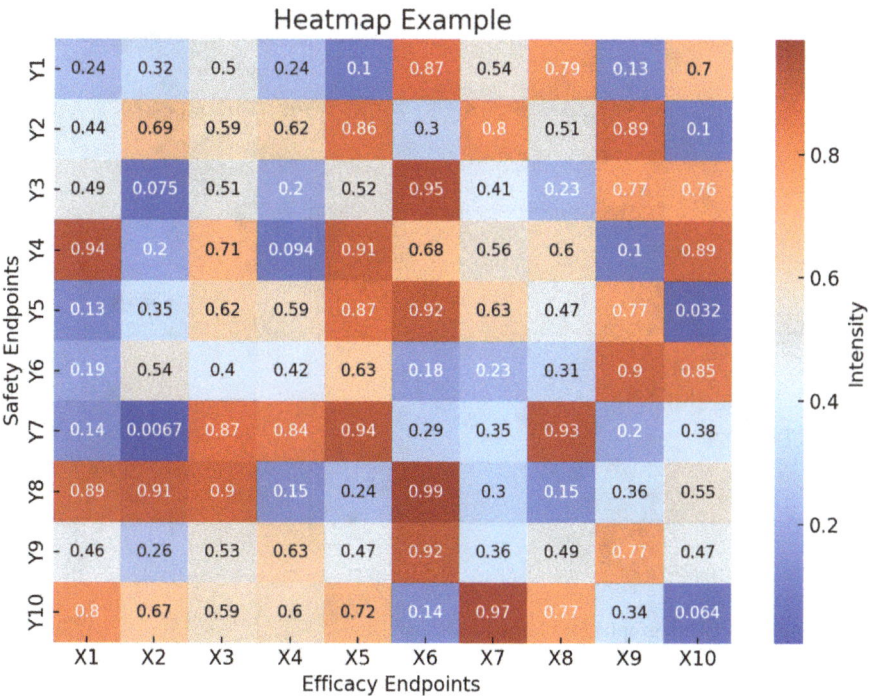

FIGURE 11.1
Heatmap visualization of dataset correlations.

for a visually intuitive representation of a correlation matrix or data relationships using color gradients to highlight intensity are a few examples.

Here is an example of a heatmap:

- The heatmap visualizes a 10×10 matrix of random values, with colors representing the intensity of values.
- Labels are included for both efficacy endpoints (*x*-axis) and safety endpoints (*y*-axis).
- The color bar provides a reference for interpreting the intensity levels (correlations).

A heatmap is a data visualization technique that represents values in a matrix or dataset using varying colors to indicate different intensities or magnitudes. Heatmaps are especially effective for visually identifying patterns, correlations, or clusters in data.

Network graphs are a powerful visualization method for hierarchical clustering and other clustering techniques, particularly when you want to illustrate relationships or connections between data points, clusters, or features. In the context of hierarchical clustering, network graphs can represent how data points are connected or grouped.

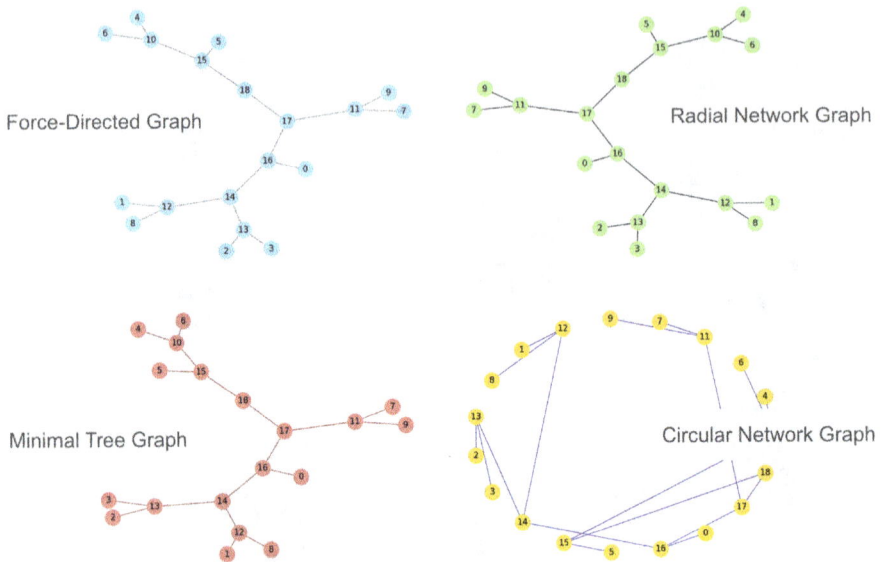

FIGURE 11.2
Four types of network graphs for hierarchical clustering.

Here are four common types of network graph visualizations for hierarchical clustering (Figure 11.2):

1. Force-directed graph: Nodes are arranged based on a force-directed
   algorithm to highlight relationships dynamically.
2. Radial network graph: Nodes are positioned in a radial layout, with
   clusters expanding outward.
3. Tree graph (minimal spanning tree): It shows the minimal connections needed to link all nodes, highlighting hierarchical structure.
4. Circular network graph: Nodes are arranged in a circular pattern for
   symmetrical visualization.

A self-organizing map (SOM) is a type of unsupervised learning algorithm
primarily used for clustering and visualizing high-dimensional data in
lower dimensions, often a 2D grid. SOMs are a neural network model, which
organizes data based on similarity without requiring labeled inputs.

Hierarchical clustering involves grouping data into a tree-like structure
based on their similarities. Visualization is a critical part of interpreting
hierarchical clustering results. Below are some commonly used visualization methods: dendrogram (Chang, 2020, p. 216), heatmap with dendrogram,
silhouette plot, and circular dendrogram.

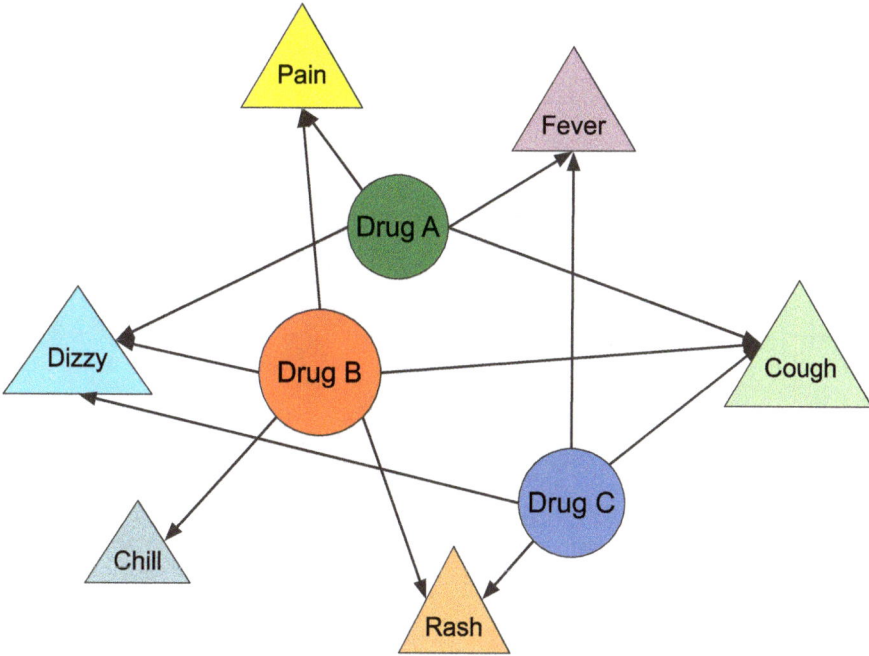

FIGURE 11.3
Drug-side effect network visualization.

In a drug-side effect network, prescription drugs are depicted as circular nodes, each uniquely colored to reflect their distinct safety profiles. Side effects are represented as triangular nodes, with varied colors highlighting the diversity of adverse effects (AEs) linked to the drugs. The size of each node corresponds to the frequency of the side effects, while the thickness (weight) of the connecting links indicates the strength of the association between a drug and a particular side effect (Figure 11.3).

In clinical trials, effectively communicating complex data through visuals is critical for delivering the right message. For instance, when a drug exhibits multifaceted effects measured across multiple endpoints at baseline and post-treatment, it is important to convey both individual and aggregate effects. While a bar chart can be used to display the drug's impact on each endpoint separately, medical professionals often seek insights into both individual effects and the overall aggregate effect.

A co-centered polygon plot (also known as a radar chart, spider plot, or web chart) is an effective visualization tool for achieving this dual purpose. In such a plot, the inner polygon represents the baseline values, while the outer polygon represents the post-treatment values (Figure 11.4). The distance from the center to a vertex of the inner polygon corresponds to the baseline value of a specific endpoint, while the distance to a vertex of the outer polygon reflects the post-treatment value for that endpoint. The space between

Co-Centered Polygon Plot: Drug Effects

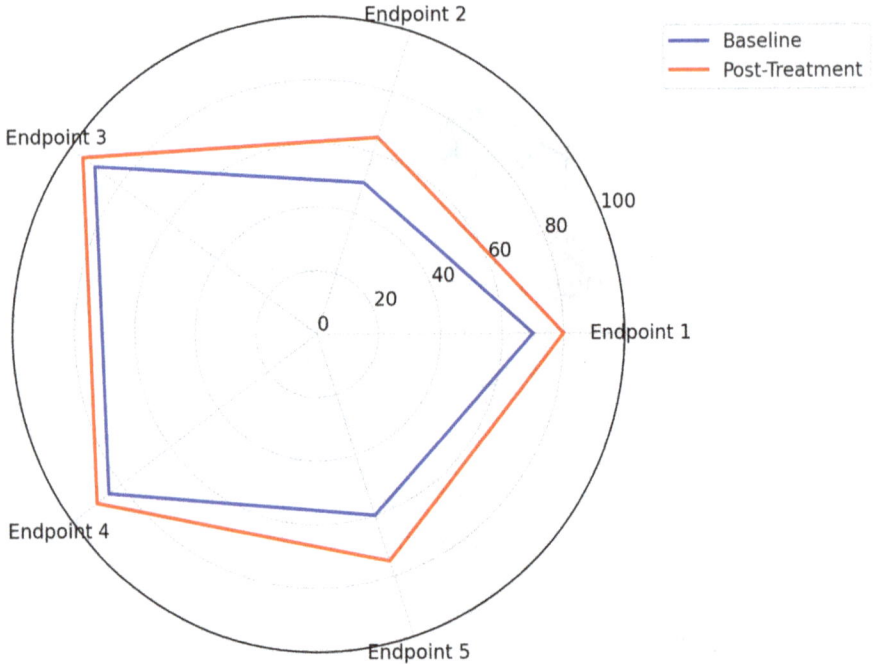

FIGURE 11.4
Multifaceted drug effects using radar charts.

the two polygons provides a clear visual representation of the drug's overall effect, offering both endpoint-specific and aggregated insights in a single, intuitive figure. Additionally, the plot effectively highlights the drug's most and least impactful endpoints.

It is crucial to use right plot to deliver the right message. Here are some tips:

The starting point of an axis can significantly influence the visual interpretation of data, either exaggerating or downplaying fluctuations, as demonstrated in Figure 11.5.

You can emphasize or de-emphasize the impact of a factor like COVID-19 by using different grouping strategies, as illustrated in Figure 11.6a and b. For example, combining deaths from heart disease and cancer (leading chronic diseases) can downplay the relative impact of COVID-19. Conversely, separating heart disease and cancer into distinct categories highlights the contribution of COVID-19 by making it appear more prominent in comparison.

We can analyze the same phenomenon (deaths) from various perspectives by using metrics such as raw death counts, death rates, changes in deaths, changes in rates, and rate-of-rate (second-order rate changes). These metrics

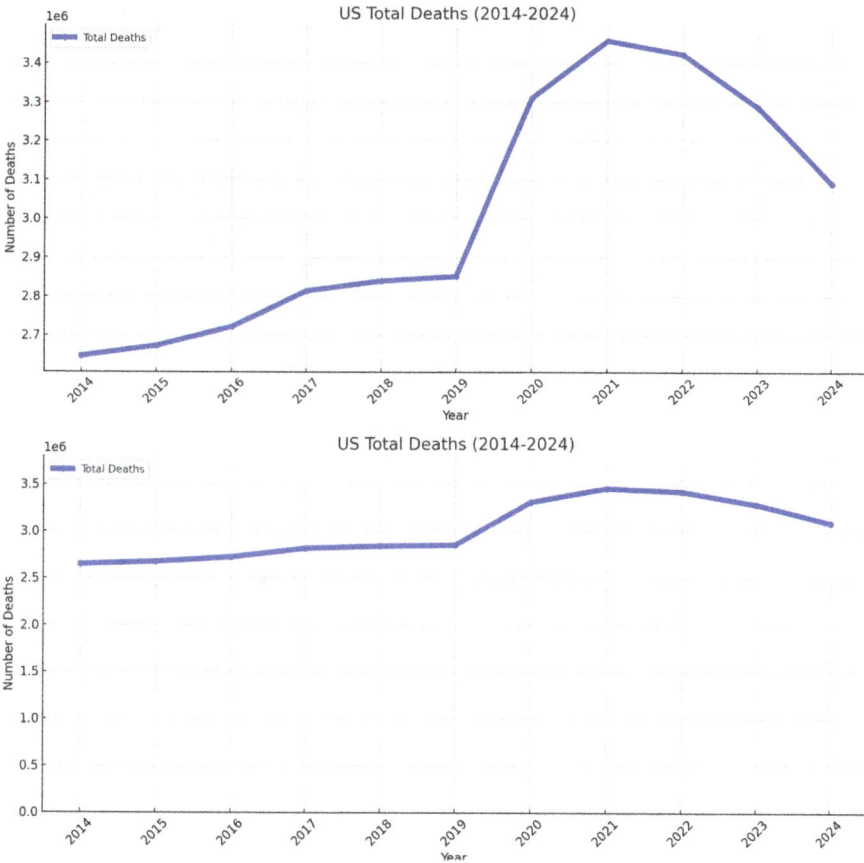

FIGURE 11.5
Scaling effects of initial data points.

offer different insights into the data. However, numerical differences between comparison groups (e.g., baseline vs. post-treatment) tend to diminish as we move to higher-order derivatives, as these changes often become negligible in practical scenarios. Therefore, if you aim to emphasize that the difference between two groups is small, a higher-order change can be used; otherwise, raw data or a lower-order change is more suitable.

Data discrepancies were identified between Figures 7.2 (expected data for the dotted lines) and 12.6 (observed data), particularly concerning the reported numbers of cancer and heart disease deaths during the COVID-19 pandemic. Figure 7.2 indicates a decline in cancer and heart disease deaths during the pandemic, consistent with the concept of a competing risk posed by COVID-19. In contrast, Figure 11.6 suggests an increase in cancer and heart disease deaths during the same period—a trend that is difficult to explain without further investigation.

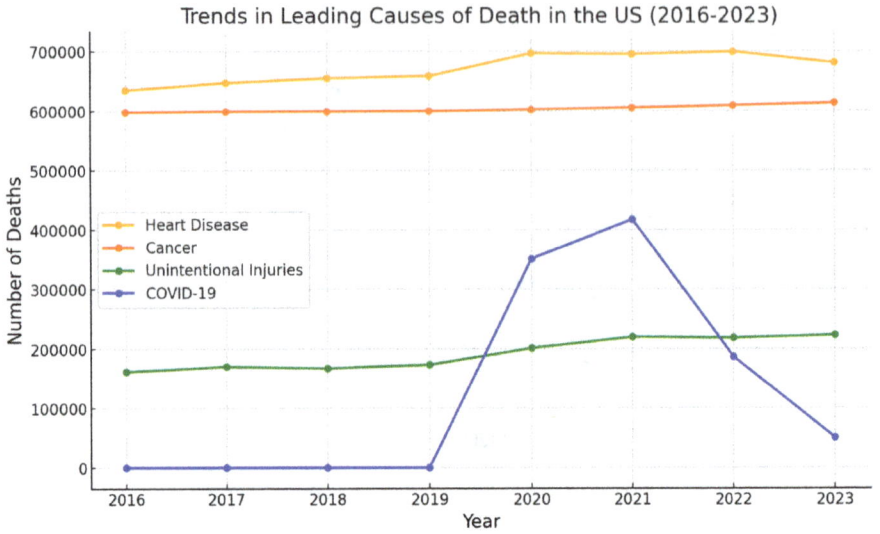

(a)

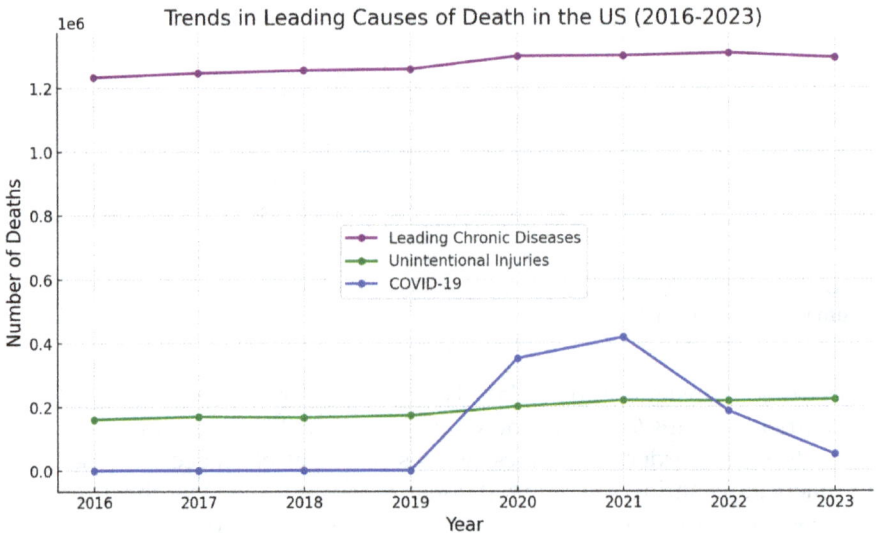

(b)

FIGURE 11.6
Impact of different groupings on analysis.

Potential explanations could include misclassification of deaths due to COVID-19 complications as cancer deaths or inaccuracies in reporting during the pandemic. However, resolving these data inconsistencies lies beyond the scope of this discussion.

11.4 Visual Analogy Creation: AI for Scientific Visualization

DALL·E is an AI-powered image generation tool integrated into ChatGPT by OpenAI. To create illustrations for a book, we can establish a consistent and visually engaging style that aligns with the book's theme. Here is how to approach it:

11.4.1 Choosing a Style

For a book on critical thinking and analogy or alike:

1. Metaphorical and conceptual style:
   - Use designs that are symbolic and concept-driven to represent abstract ideas (e.g., recursion, paradox, and logical frameworks).
   - Keep the visuals minimalistic but impactful, focusing on clarity and elegance.
2. Cartoon style:
   - A cartoon or playful style works well for making complex ideas approachable and engaging.
   - Use clean lines and simple colors, balancing humor and thoughtfulness.
3. Modern and abstract style:
   - For a sophisticated, AI-focused theme, consider a geometric or abstract style with soft gradients and clean shapes.
   - This style complements high-tech and intellectual topics.

11.4.2 Using DALL·E for Consistency

When creating illustrations using DALL·E, specify stylistic elements clearly in your prompts:

1. Key phrases for conceptual and metaphorical illustrations:
   - Use terms like "metaphorical," "symbolic," "minimalist," and "thought-provoking."
   - Include objects or scenes that represent ideas clearly (e.g., "a human brain holding infinity symbols").
2. Key phrases for cartoon style:
   - Add "cartoon-style," "playful," or "whimsical" for engaging designs.
   - Specify "clean lines," "vivid colors," or "minimalistic background."

3. Key phrases for abstract style:
   - Use "modern," "abstract," or "geometric."
   - Describe light and color effects, such as "soft gradients," "monochromatic palette," or "high-tech feel."

11.4.3 Prompt Writing Tips

- Focus on the key idea: Clearly describe the metaphor or analogy you want to illustrate.
  - Example: "A cartoon-style illustration of a person holding a glowing question mark, symbolizing curiosity and inquiry."
- Specify visual elements:
  - Example: "A metaphorical illustration of recursion: a head inside a head inside another head, minimalistic background."
- Use adjectives for style: Words like "minimalist," "playful," "vibrant," or "symbolic" help guide the visual tone.
  - Example: "A vibrant cartoon-style illustration showing gears inside a brain, representing critical thinking."
- Be concise: Limit prompts to essential elements to avoid cluttered designs.
  - Example: "An abstract illustration of a human hand holding a glowing infinity symbol, set against a futuristic gradient background."

11.4.4 Sample Prompts

1. For critical thinking:
   - "A metaphorical illustration of a brain with lightbulbs and gears inside, symbolizing critical thinking, minimalist design."
2. For creative analogy:
   - "A whimsical cartoon of two puzzle pieces being connected by a person, symbolizing creative analogy."
3. For AI themes:
   - "An abstract illustration of a neural network interwoven with glowing human figures, representing the synergy between AI and human creativity."

11.4.5 Experiment and Refine

1. Start with a few prompts to establish a consistent style.

2. Adjust based on feedback—refine the level of detail, colors, and tone to align with your book's themes.

3. Use feedback loops (test multiple prompts and pick the best results) to ensure cohesiveness across all illustrations.

To give you a visual understanding of how AI prompts work, we have included several examples showcasing the effects of different prompts and artistic styles. These styles include minimalist cartoon, scientific illustration, colored pencil drawing, minimalist, playful, educational, whimsical, and surreal. Below are the image prompts along with the corresponding figures generated by OpenAI's DALL·E:

- **Figure 10.1 Prompt**: Create a drawing of the metaphor of the self-reference paradox in a minimalist cartoon style.

- **Figure 10.2 Prompt**: Draw a scientific illustration of "The Self-Inclusive Mind Enabled Self-Awareness."

- **Figure 20.1 Prompt**: Draw a scientific illustration of "Socrates versus Confucius."

- **Figure 13.1 Prompt**: Draw an image of an adjustable-length seesaw to illustrate quantum entanglement in a color pencil drawing, educational style.

- **Figure 13.3 Prompt**: Create a cartoon illustrating a 2D creature puzzled by a 3D sphere passing through its 2D plane, highlighting the dimensional contrast and the creature's limited understanding in a minimalist, playful, educational style.

If you want a **consistent style** across a group of images, specify the same style. For example, as shown in **Figure 13.2**, the three images maintain remarkable consistency in style when using a whimsical approach:

- **Figure 13.2a Prompt**: Create a cartoon metaphor: A summer mosquito lives its brief life in the warmth of the season, entirely incapable of comprehending what winter might feel like, whimsical style.

- **Figure 13.2b Prompt**: Draw a cartoon of a deaf person trying to hear music, enhancing the sense of deafness, whimsical style.

- **Figure 13.2c Prompt**: Create a metaphorical image of a blind person living in a colorful world, whimsical style.

Whimsical and surreal styles are especially effective for illustrating abstract concepts, such as in the following prompt:

- **Figure 13.4 Prompt**: Create an image in whimsical and surreal styles to depict the metaphor of an adjustable-length seesaw for quantum entanglement and cats passing through a portal.

Note that AI drawing tools like DALL.E are typically not ideal for generating exact text. The text in Figures 13.3 and 13.4 has been manually added. Additionally, the same prompt can produce different images each time due to inherent randomness and varying conversation history with the AI tool.

Even though the final prompts are not very complex, the long iteration process is unavoidable. With this foundational knowledge, you can begin creating images with purpose. Continuous practice and improvement will help you become a skilled prompt engineer.

11.5 Summary: Creative Data Visualization and Its Role in Communication

This chapter highlights the transformative power of creative data visualization as a tool for bridging abstract data and intuitive understanding. By leveraging analogies and storytelling, effective visualizations convey complex insights in ways that are engaging, memorable, and actionable. The chapter explores the balance between creativity and accuracy, emphasizing the need to inform without misleading. It also addresses common challenges, such as oversimplification, unclear narratives, and audience misalignment, offering strategies for enhancing both clarity and emotional impact.

Key Takeaways

1. Balancing creativity and integrity: The chapter emphasizes the importance of maintaining accuracy while using creative visuals to engage audiences. It critiques misleading practices, such as inappropriate scaling, different grouping, or excessive data density, that distort insights.

2. Audience-centric design: By aligning visualizations with the needs and cognitive preferences of viewers, the chapter demonstrates how thoughtful design fosters comprehension and decision-making.

3. Critical use of visualizations: Through examples of appropriate graph selection and grouping strategies, the chapter encourages readers to critically evaluate visual representations for transparency and relevance.

Exercise

1. Explain the importance of clarity, simplicity, and aesthetic appeal in data visualization. Provide an example of a well-designed chart and describe how it embodies these principles.

2. Match the following data types (binary, nominal, ratio, and time-series) to the most appropriate visualization type (e.g., bar chart, scatter plot, line chart, and heatmap). Justify your choices.

3. Design a visualization to compare vaccination rates across countries. Specify the type of graph you would use, the data to include, and how you would highlight key insights.

4. Create or analyze a heatmap of correlation values for different variables in a dataset (e.g., health metrics or economic indicators). Interpret the patterns and provide actionable insights.

5. Create an analogy-based data visualization. For instance, use a spiral plot to represent seasonal sales trends or a radial chart to compare different departments' performance in an organization.

6. Propose a creative way to visualize hierarchical data using one of the techniques mentioned in the chapter (e.g., radial network graph or circular dendrogram). Explain your rationale and potential use case.

7. Discuss how visualization choices (e.g., starting axis points and grouping strategies) can influence perception. How can designers balance impact with integrity?

8. Debate the value of interactive visualizations compared to static ones. When might interactivity enhance understanding, and when could it overwhelm the viewer?

9. Explore how AI tools like DALL·E, MidJourney, Runway ML, or Tableau can enhance creative data visualization. Design a prompt for an AI tool to generate a conceptual data illustration, ensuring it aligns with a specific message or theme.

10. Propose a dynamic visualization for tracking real-time metrics, such as website traffic or stock prices. Describe how this visualization could help users identify trends and make decisions.

11. From the drug-side effect network (Figure 11.3), we can develop a software tool to facilitate the exploration of this information. For instance, the user interface could feature nodes representing drugs and AEs, each accompanied by corresponding visual elements. When a user selects a specific drug, its associated AEs will be displayed as neighboring nodes in the network, along with descriptive information presented in a side panel. Similarly, clicking on an AE node will highlight the related drugs as neighboring nodes and display additional relevant details.

As a team, discuss the design and functionality of this tool and plan its development using software such as Swift in Xcode. This exercise is proposed as a term project to combine technical, design, and collaborative skills in building an interactive and educational application.

Student Project Instruction Sheet

Developing an Interactive Drug-Side Effect Exploration Tool

Context

Using the drug-side effect network (e.g., Figure 11.3), your goal is to create an interactive software tool that allows users to explore relationships between prescription drugs and their associated AEs. The tool will visually represent the network and provide intuitive ways to access detailed information about drugs and AEs.

Objectives

- Design a user-friendly interface to visualize the drug-side effect network.
- Implement functionality to display detailed information when a user interacts with specific nodes (drug or AE).
- Utilize a software development framework such as **Swift** (via Xcode) to build the tool.

Features to Include

1. **Node interaction:**
   - **Drug nodes:** When clicked, display all associated AE nodes as neighbors and provide textual descriptions in a side panel.
   - **AE nodes:** When clicked, display all associated drug nodes and their detailed safety profiles or other relevant information.
2. **Dynamic visualizations:**
   - Represent drugs as circular nodes and AEs as triangular nodes, with intuitive sizing and coloring (as in the network diagram).
   - Use animations to highlight connected nodes upon selection.
3. **Search and filter:**
   - Add a search bar to allow users to locate specific drugs or AEs easily.
   - Provide filtering options, such as by severity of AEs or drug class.
4. **Information panel:**
   - Include a dedicated section for detailed descriptions of the selected drug or AE.
   - Allow for expansion or detailed views, including references or links to additional resources.

5. **Customization:**
   - Enable users to adjust visualization parameters (e.g., node sizes, link weights, or layout styles).

Development Process

1. **Team brainstorming:**
   - Discuss and finalize the tool's core features and design.
   - Draft wireframes and mockups for the user interface (UI).
   - Decide on data structure and representation (e.g., adjacency list and graph database).
2. **Implementation:**
   - Use **Swift** in Xcode to build the application.
   - Integrate a graph visualization library (e.g., **SwiftGraph**, **GraphKit**, or custom rendering).
   - Implement event-driven programming for node interactions.
3. **Testing and feedback:**
   - Conduct usability testing with peers to refine the interface and functionality.
   - Address bugs and improve performance, especially for larger networks.
4. **Final deliverable:**
   - Present a working prototype of the tool.
   - Include documentation covering features, design decisions, and potential extensions.

Deliverables

- A functional interactive tool showcasing the drug-side effect network.
- A presentation discussing your team's design and development process.
- A report detailing the tool's features, implementation, and future scope.

Suggested Software

- **Swift** with **Xcode**: Ideal for developing iOS-based interactive applications.
- **Graph libraries:** Libraries like **SwiftGraph** for implementing graph-based functionalities.

- **Data sources:** Use CSV files or JSON data to feed the network structure into the tool.

Term Project Goals

This exercise encourages hands-on experience in software development, team collaboration, and applying data visualization techniques. Discuss creative ideas within your team and aim for a functional tool that bridges science and technology effectively.

12

From Brownian Motion to Unified Models: Bridging Scales and Disciplines

If multiple problems share the same equation, one solution applies to all, showcasing analogy's rigor.

12.1 Stochastic Process of Particle Random Movement

Brownian motion is a phenomenon observed as the random, erratic movement of particles suspended in a fluid (liquid or gas) due to collisions with the molecules of the fluid. Named after botanist Robert Brown, who observed pollen grains moving unpredictably in water in 1827, this motion provided one of the earliest forms of evidence for the existence of atoms and molecules.

Brownian motion is highly irregular and unpredictable, with each particle's movement being independent of its past movements. There is no fixed direction or magnitude; the particles move continuously and randomly due to incessant collisions with the surrounding fluid molecules.

In mathematical terms, Brownian motion (Figure 12.1) is modeled by a Wiener process $W(t)$, which has the following properties:

1. Initial value: $W(t) = 0$, meaning the motion starts at the origin.
2. Independent increments: The changes in position over non-overlapping time intervals are independent.
3. Normal distribution of increments: For any time t, the increment $W(t) - W(s)$ (where $t > s$) follows a normal distribution with mean zero and variance proportional to the time interval $t - s$: $W(t) - W(s) \sim N(0, t - s)$.
4. Continuous paths: The path of $W(t)$ is continuous over time, although it is nowhere differentiable, meaning it's highly irregular.

Brownian motion exhibits a fascinating relationship with self-similarity, which refers to the property where a process or structure appears similar at different scales (Chapter 10). For Brownian motion, this means that if you zoom in on a segment of a Brownian path, it looks statistically similar to the entire path, regardless of the scale.

DOI: 10.1201/9781003630081-12

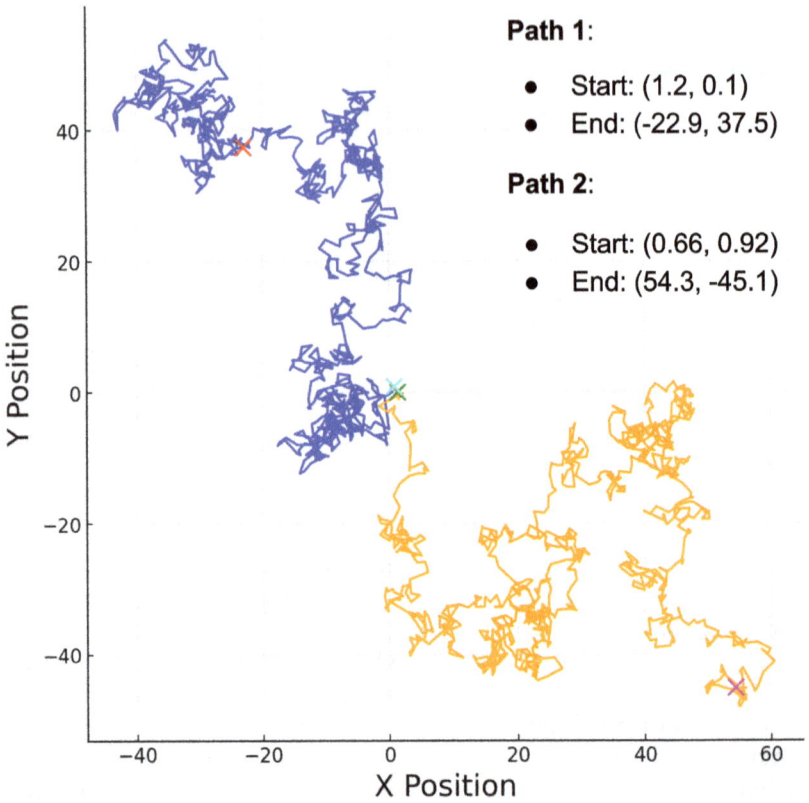

FIGURE 12.1
Simulated 2D Brownian motion paths.

Scale invariance: Brownian motion is a fractal process with scale invariance, meaning that its statistical properties remain the same when the time and spatial scales are adjusted proportionally. Mathematically, if you rescale time and space by a factor of a, the trajectory of Brownian motion $W(at) = \sqrt{a}W(t)$ in distribution. This means that randomly walking 2 units away will take 4 times as long as walking 1 unit away.

If multiple problems share the same underlying equation, one solution can be applied across disciplines, demonstrating the power of analogy. In physics and chemistry, Einstein's work on Brownian motion provided a foundation for atomic theory, enabling calculations of Avogadro's number and supporting molecular kinetic theory. In finance, it underpins the Black–Scholes model for option pricing, where asset prices are treated as geometric Brownian motion to capture market fluctuations.

Similarly, in biology, Brownian motion explains the diffusion of nutrients and molecules within cells, essential for processes like signal transduction and molecular binding. In medicine and statistics, adaptive clinical trial designs use Brownian motion to set stopping boundaries and control error rates. Even in engineering and mathematics, many problems are analogs of Brownian motion and governed by the Wiener process, underscoring the versatility of this framework.

Let us take temperature as an example to elaborate this further. We know the temperature at location x is the average speed of particle random movement (Brownian motion),

$$E\left[u\left(B(t)\right)\right] = u(x).$$

The expectation $E\left[u\left(B(t)\right)\right]$ is computed over the ensemble of possible paths that the Brownian particle can take by time t, weighted by their probabilities.

A function $u(x)$ is harmonic if it satisfies the Laplace equation:

$$\nabla^2 u = 0.$$

where the Laplace operator $\nabla^2 = \dfrac{\partial^2}{\partial x^2} + \dfrac{\partial^2}{\partial y^2} + \dfrac{\partial^2}{\partial z^2}.$

A key property of harmonic functions is the mean value property: the value of $u(x)$ at a point is equal to the average value of u on any sphere centered at that point.

Brownian motion often involves stopping times, which are the times at which the particle first exits a region: The probability of the first exiting and the $u(x)$. In relationship to the boundary values have broach applications, including stopping boundaries for adaptive clinical trials, general threshold regressions, and boundary problems of partial differential or integral equations in mathematics and engineering.

12.2 Adaptive Clinical Trial

An adaptive clinical trial is a type of clinical study design that allows for interim analyses of data at predefined points during the trial. These interim checks enable researchers to determine whether the trial should be stopped early for reasons such as demonstrated efficacy, futility, or safety concerns. By incorporating these analyses, adaptive designs improve efficiency, reduce costs, and minimize the number of participants exposed to potentially less effective treatments.

Statistical methods are employed to control the overall Type I error rate (false positives) across multiple interim analyses. This is typically achieved using stopping boundaries, which define thresholds for early termination or continuation of the trial. Adaptive trials are widely used in drug development and other medical research to accelerate the evaluation of treatments while maintaining robust scientific rigor.

Brownian motion can have another for random motions, called discrete Brownian motion, which occurs only at discrete time intervals. Discrete Brownian motion serves as a valuable analogy in adaptive clinical trial design, offering a mathematical framework that underpins the statistical properties of repeated testing over time. This connection stems from Brownian motion's predictable, continuous, and independent increments, which can model the cumulative progress of evidence (such as test statistics) as data accumulates in a clinical trial. Here is how this analogy works and why it is particularly useful in this context.

12.2.1 Sequential Testing and Brownian Motion

In adaptive clinical trials, interim analyses are conducted at predetermined points to monitor accumulating evidence on treatment effectiveness. The purpose of these analyses is to determine whether the trial can be stopped early for efficacy, futility, or safety. However, performing multiple interim tests increases the risk of false positives, so careful statistical control is essential.

Brownian motion provides a natural analogy for modeling these repeated interim looks:

- Cumulative evidence: Just as Brownian motion describes a path that accumulates increments over time, the cumulative test statistics in a trial accumulate as data from each group becomes available.
- Predictability of variability: Brownian motion has well-defined properties for the variability of a process over time, allowing us to set boundaries for cumulative test statistics at each interim point in a trial.

12.2.2 Discrete Brownian Motion as a Model for Test Statistics

In the context of adaptive trials, the test statistic's cumulative path can be modeled by a one-dimensional (bidirectional) discrete Brownian motion, where the observed statistic at each interim analysis is treated as an incrementally progressing particle random movement. This framework allows statisticians to:

- Define boundaries for early stopping: With Brownian motion, it is possible to establish stopping boundaries that control the type I error rate (false positive rate) across multiple interim analyses. This is analogous to setting boundaries for the path of the Wiener process to keep the cumulative evidence within a specified range.

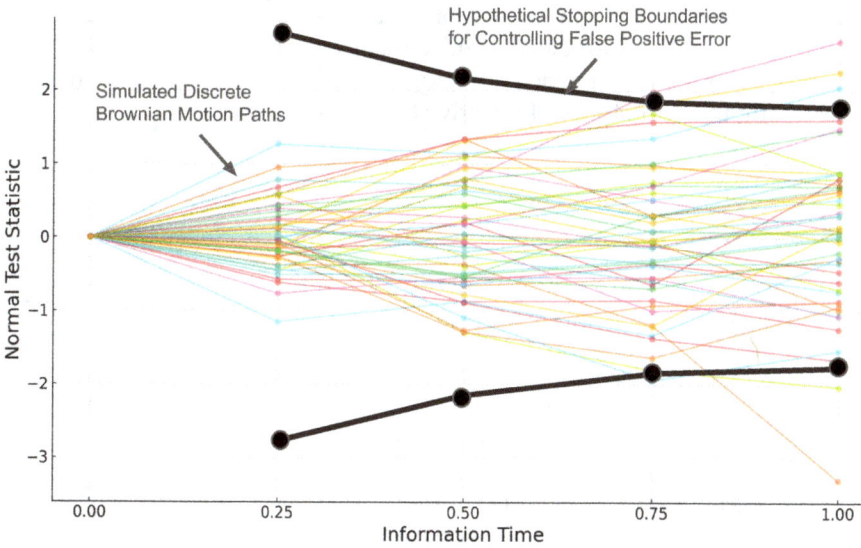

FIGURE 12.2
Adaptive clinical trial stopping rules illustrated with Brownian motion.

- Calculate probabilities of crossing boundaries (the first exit time): Brownian motion's properties enable precise calculations for the probability of the path crossing certain boundaries by each analysis point. This helps control the overall significance level across multiple tests.

For example, if the cumulative evidence (analogous to the path of Brownian motion) crosses a predefined upper boundary, it may indicate strong evidence of treatment efficacy, warranting early trial termination (Figure 12.2). The same analogy can be applied to the so-called Lan–DeMets spending function approach for adaptive clinical trial design. The spending function approach allows us to have as many interim analyses as we want and whenever we want.

Lan and Demets (1994) consider an adaptive trial with K analyses with associated test statistics $\{Z_1, Z_2, ..., Z_K\}$, where for large sample size,

$$Z_K \sim N\left(\theta\sqrt{I_k}, 1\right), \quad 1, ..., K; \quad \text{Cov}\left(Z_p, Z_q\right), \quad 1 \le p \le q \le K.$$

Here, the information level (information time) I_k is a function of sample size and variance, e.g., for comparing two means in a balanced design, $I_k = \dfrac{N_k}{2\sigma^2}$, where N_k is the cumulative sample size per group up to stage k, and σ^2 is the common variance.

Adaptive designs have been instrumental in the expedited approval of many prescription drugs by allowing for interim analyses that can lead to early trial termination due to demonstrated efficacy or futility. Adaptive designs enhance trial efficiency and ethical considerations by potentially reducing the number of participants exposed to less effective treatments and accelerating the availability of effective therapies to patients.

12.3 Threshold Regression for the Wiener Process of Hidden Biomarkers

Brownian motion processes can be applied far beyond their traditional use in adaptive clinical trial designs. Notably, the use of Brownian motion in adaptive trials represents a specific application within the broader framework of threshold regression.

Threshold regression (Lee and Whitmore, 2006; Lee, Chang, and Whitmore, 2008; Chang, 2019, pp. 174–1184) is a powerful statistical modeling approach often employed to analyze time-to-event data, particularly in situations where the occurrence of an event is triggered by an underlying process reaching an unobservable threshold. This framework is especially valuable in studying hidden biomarkers—unmeasurable biological variables that influence the timing or likelihood of critical events, such as disease onset, progression, or recovery.

Note that while tumor size can often be directly measured, it may still be considered "unobservable" in specific modeling contexts where continuous observation is impractical or unavailable, subclinical or dynamic changes occur between measurements, statistical models are needed to infer its progression, and the "unobservable" nature in such cases reflects the limitations of real-time or exact measurement, not the inherent invisibility of the tumor itself.

12.3.1 Key Concepts of Threshold Regression

1. Hidden biomarker: Represents an unobservable biological variable (e.g., tumor size, immune response level, or molecular concentration) that progresses over time and affects the risk of an event (e.g., disease onset, relapse, or recovery). This biomarker evolves according to a stochastic process or Wiener process.

2. Threshold: A critical level or boundary in the hidden biomarker process. The event of interest (e.g., symptom appearance, diagnosis, and failure) occurs when the hidden biomarker crosses this threshold.

3. Stochastic process: The biomarker's trajectory is often modeled as a Brownian motion (Wiener process), or another stochastic process, potentially with drift, reflecting its random but directional evolution.

4. Time-to-event: The primary variable of interest is the time at which the hidden biomarker crosses the threshold.

12.3.2 Threshold Regression Model Structure

1. Model assumptions: The hidden biomarker follows a stochastic process $X(t)$, evolving over time t. The event occurs when $X(t)$ reaches or exceeds a pre-specified threshold τ:

$$T = \inf\{t : X(t) \geq \tau\},$$

where the threshold τ may be constant or vary based on covariates or individual characteristics.

2. Covariates: Observable covariates Z influence the biomarker's progression, threshold level, or both. For example: the drift term in the stochastic process (mean trajectory), the volatility term (variability in progression), and the threshold level itself.

3. Likelihood function: Derived from the stochastic process governing $X(t)$, incorporating survival data and censoring.

4. Output: Estimates of parameters governing the biomarker's dynamics and threshold characteristics and predicted time-to-event or risk probabilities for new observations.

12.3.3 Applications to Hidden Biomarkers

Threshold regression is particularly effective in scenarios where biomarkers are not directly observable or are measured only intermittently. In cancer research, for example, a hidden biomarker such as tumor size or genetic mutation load may grow stochastically over time. An event, such as cancer diagnosis, recurrence, or metastasis, occurs when the tumor crosses a detectable threshold, making this approach invaluable for modeling disease progression.

In immunology, threshold regression can be used to study hidden biomarkers like antibody concentration or immune response. Here, an event such as infection resistance or disease onset arises when the immune system fails to maintain these biomarkers above a critical threshold, illustrating how immune response evolves over time.

For cardiovascular health, hidden biomarkers like atherosclerotic plaque size or blood pressure variability can be modeled using threshold regression. Events such as heart attacks or strokes occur when these biomarkers grow stochastically and surpass critical levels, offering insights into the underlying mechanisms of cardiovascular events.

Neurodegenerative diseases also benefit from this approach, where hidden biomarkers such as neuronal degeneration or amyloid-beta levels play a key role. Clinical symptoms, including cognitive decline or dementia diagnosis, manifest when these biomarkers progress beyond functional thresholds.

In drug development, hidden biomarkers like pharmacokinetic or pharmacodynamic responses to treatment are crucial. Therapeutic effects or adverse reactions occur when drug concentrations in the body cross specific thresholds. Threshold regression models provide a framework for understanding how drug levels evolve and their impact on efficacy or toxicity.

By capturing the dynamics of hidden biomarkers and linking them to observable clinical events, threshold regression serves as a powerful tool across these diverse applications.

12.3.4 Advantages of Threshold Regression

1. Biological interpretability: Directly connects the event to an underlying biological process, offering mechanistic insights.

2. Handles latent variables: Accommodates unobservable biomarkers, crucial in fields like oncology or neurology where key processes are not easily measured.

3. Incorporates covariates: Allows covariates to influence the stochastic process and threshold, making the model flexible and patient-specific.

4. Censoring and survival data: Naturally integrates censored data, a common feature in clinical and longitudinal studies.

12.3.5 Challenges and Limitations

1. Model complexity: Requires advanced statistical techniques to estimate parameters, especially for high-dimensional or non-linear covariate effects.

2. Stochastic process specification: Choosing the correct stochastic process (e.g., Brownian motion, Ornstein–Uhlenbeck process) is critical but non-trivial.

3. Threshold identification: Determining the threshold level or its dependence on covariates can be challenging without prior biological knowledge.

4. Data limitations: Sparse or noisy data on observable proxies for the hidden biomarker can reduce model accuracy.

In conclusion, threshold regression is a powerful tool for understanding time-to-event processes driven by hidden biomarkers. It bridges stochastic process modeling with survival analysis, offering deep insights into unobservable biological mechanisms. By capturing the interplay between biomarker dynamics and threshold events, it has wide applicability in biomedical research, personalized medicine, and drug development. Its continued evolution promises to improve predictive accuracy and mechanistic understanding in complex, data-rich environments.

12.4 Random Walk Method for Partial Differential Equations

In Section 12.1, we described how the macroscopic behavior of small particles undergoing discrete random motion at the microscopic level can be modeled approximately using Laplace partial differential equations (PDEs) of a continuous field. In this section, we reverse the perspective: starting with PDEs, we use computer-simulated "random walks" to numerically solve these equations. Random walk methods provide a powerful algorithmic approach for addressing complex problems in mathematics and engineering.

A random walk forms a stochastic process that consists of a sequence of discrete steps, each determined randomly. Movement occurs in steps, usually over a grid or lattice. Each step is random and often independent of previous steps (e.g., moving left, right, up, or down).

A random walk serves as a discrete approximation of Brownian motion. The connection becomes exact in the limit as time and space intervals shrink to zero, with the step size scaled appropriately. This relationship bridges discrete and continuous modeling.

Differential equations are equations that involve an unknown function and its derivatives. They are fundamental in describing various physical phenomena in engineering, physics, economics, biology, and more.

Diffusion equation is a differential equation to describe diffusion processes such as heat distribution, electrostatic potential, and incompressible fluid flow. It can be written as:

$$\frac{\partial u(r,t)}{\partial t} = D\nabla^2 u(r,t),$$

where r is the position vector and D is a constant.

The equation becomes Laplace equation for steady state diffusion: $\nabla^2 u(r) = 0$.

To solve this equation, we require specified boundary conditions on the domain's edges. These conditions can involve:

- Dirichlet boundary conditions: specifying the value of $u(r)$ on the boundary.

- Neumann boundary conditions: specifying the derivative of $u(r)$ normal to the boundary.

By applying these boundary conditions, we can determine a unique solution for $u(r)$ within the domain.

The Laplace equation is widely used in modeling seepage through earth dams to describe the movement of water through porous media. This application relies on the fact that seepage flow is governed by a potential field that satisfies the Laplace equation, where $u(r)$ is the hydraulic head (a measure of the potential energy of the water at a point), and r is the location within the dam. This equation models the steady-state condition, meaning that the flow and hydraulic head distribution within the dam are constant over time.

The Laplace equation implies that the hydraulic potential is a harmonic function, meaning that at any point within the flow domain, the potential is the average of the potentials in the surrounding area. This behavior describes the steady-state distribution of hydraulic potential under conditions where water infiltrates through the dam from the upstream side to the downstream side.

Random walk interpretation for steady-state diffusion: The value of the solution $u(r_0)$ at a point (r_0) can be estimated by averaging the boundary values reached by random paths (walks) starting at (r_0). From this notion, the random-walk method is developed to solve Laplace's equation and Poisson's equation via Monte Carlo simulation–simulated random walk on a computer. In other words, if a drunk, starting from r_0, randomly walks until he reaches the boundary of a domain with known value of $u(r_b)$ and repeat the process many times, then the average of the boundary values $u(r_b)$ is the value of $u(r_0)$ to be solved.

12.5 Dirac's Delta and Green's Functions

12.5.1 Unification of Discrete and Continuous Probability Distributions

The Dirac delta function plays a significant role in advanced mathematics, physics, and engineering. Although commonly referred to as a "function," it is more accurately described as a generalized function, as it does not conform to the classical definition of functions. Its primary purpose is to model idealized impulses or point sources, making it a versatile and powerful tool across various disciplines.

The Dirac delta function, denoted $\delta(x)$, has the following defining properties:

1. Sifting property:

$$\int_{-\infty}^{\infty} \delta(x-a)f(x) = f(a),$$

for any test function $f(x)$. This property allows $\delta(x-a)$ to "extract" the value of $f(x)$ at $x = a$.

2. Localization:

$$\delta(x) = \infty, \quad \text{if } x = a, \quad 0 \text{ otherwise.}$$

However, $\delta(x)$ is not a standard function because its value at $x = a$ is undefined in the classical sense. Its meaningfulness arises only under an integral.

The Dirac delta function can be used to define discrete probability distributions as the examples shown below:

Discrete distributions: For a discrete random variable X taking values $x_1, x_2, \ldots, x_n$ with probabilities $p_1, p_2, \ldots, p_n$, the probability distribution function (PDF) can be expressed using the Dirac delta function $\delta(x - x_i)$:

$$f(x) = \sum_{i=1}^{n} p_i \delta(x - x_i).$$

If the PDF is Dirac's function $\delta(x - x_i)$, the cumulative distribution function is the Heaviside step function

$$H(x - x_i) = 0 \quad \text{if } x < x_i \quad \text{and} \quad 1, \text{otherwise.}$$

Mixed distributions: For random variables that are a mixture of discrete and continuous components, the PDF can combine both forms:

$$f(x) = \sum_{i=1}^{n} p_i \delta(x - x_i) + f_c(x),$$

where $f_c(x)$ is the continuous part of the distribution, and the discrete part is represented by delta functions.

The normalization requires the total probability must integrate to 1:

$$\int_{-\infty}^{\infty} f(x)dx = \sum_{i=1}^{n} p_i + \int_{-\infty}^{\infty} f_c(x)dx = 1.$$

12.5.2 Green's Function

Green's function is a powerful tool in mathematics, physics, and engineering, used primarily to solve linear differential equations and integration equations subject to specific boundary or initial conditions. It acts as a tool to transform complex problems into more manageable forms, leveraging the properties of linear systems.

Green function $G(r,r')$ can be used to link one function $f(r')$ at one location (such as a point force applied at location r') to another function at another location $u(r)$ (such as a strain or deformation at location y). Mathematically, it is written in a differential form:

$$LG(r,r') = \delta(r-r'),$$

where L is a linear differential operator such as a Laplace operator.

With Green's function, the value of $u(r)$ at point r can be written in an integral form:

$$u(r) = \int_B G(r,r) f(r') dr'.$$

When $f(r')$ and $u(r')$ are given on the boundary B, the $u(r)$ can be easily obtained through a numerical integral as in the boundary element method.

As a simple example, the Green functions for conductivity of isotropic materials are:

$$G(r,r') = \frac{1}{4\pi\sigma|r-r'|} \text{ for a 3D problem,}$$

$$\text{and } G(r,r') = \frac{\ln|r-r'|}{2\pi\sigma} \text{ for a 2D problem.}$$

where constant σ is conductivity for electric field problems or permeability for seepage problems.

12.5.3 Green's Function Have the Following Key Properties

1. Linearity: Green's functions apply only to linear systems, as the superposition principle is essential for their derivation.
2. Boundary conditions: Green's functions may or may not be constructed to satisfy the boundary or initial conditions of the specific problem.
3. Impulse response: Green's function represents the response of the system to a localized unit impulse.

Green's functions have a wide range of applications across physics, engineering, and mathematics. In physics, they are used to determine the potential due to a charge distribution in electrostatics, solve Schrödinger's equation for potential problems in quantum mechanics, and understand wave propagation in various media.

Using Green's function and Dirac-delta function, we can derive the Lippmann–Schwinger equation from the time-independent Schrödinger equation:

$$\psi(r) = \psi_0(r) + \frac{\hbar^2}{2m} \int G(r, r') V(r') \psi(r') dr',$$

Here $\psi_0(r)$ is the free-particle wavefunction and $V(r')$ is the potential.

The Lippmann–Schwinger equation is a cornerstone of quantum mechanics, particularly in the study of scattering theory. Its generalized form has also been applied in classical elastic mechanics to address problems involving glandular materials and seepage, utilizing both isotropic and anisotropic Green's functions (Chang and Chang, 1995).

In this context, the Dirac delta function and Green's functions emerge as complementary tools for solving complex problems in various fields. The Dirac delta function serves as a bridge between discrete and continuous domains, modeling impulses and enabling the formulation of mixed distributions. Green's functions build upon this foundation, offering elegant solutions to partial differential equations and boundary value problems by capturing system responses to localized impulses.

12.6 Summary: Unified Models through Analogies and Brownian Motion

This chapter explores the profound connections between Brownian motion, mathematical modeling, and their diverse applications across physics, biology, medicine, and engineering. By linking stochastic processes to real-world phenomena, the chapter demonstrates how unified models and analogies streamline understanding across disciplines. Examples include Brownian motion in diffusion processes, its use in clinical trial designs, and its connections to quantum mechanics through analogies facilitated by Dirac delta function and Green's function. Together, the Dirac delta function and Green's functions exemplify the deep interconnections between mathematical theory and practical applications. They provide a unified framework for addressing diverse challenges in science and engineering, from quantum mechanics to continuum mechanics and beyond.

Key Takeaways

1. Unified frameworks: The chapter emphasizes how shared mathematical equations (e.g., those modeling Brownian motion) provide solutions applicable to multiple fields, fostering interdisciplinary insights.

2. Brownian motion as a universal analogy: Described as a random walk, Brownian motion is likened to unpredictable paths in nature,

such as the diffusion of particles, quantum mechanics, or financial market fluctuations, fostering a relatable understanding.

3. Visual analogies in modeling: The analogy of random walks simulating diffusion highlights how iterative testing mirrors natural systems, while threshold regressions offer metaphors for decision-making processes, such as crossing critical boundaries.

4. Interdisciplinary analogies: Tools like Green's functions and the Dirac delta function illustrate how mathematical abstractions reveal shared principles in diverse systems, from quantum mechanics to hydrology.

This chapter exemplifies the power of critical thinking and creative analogies, showcasing how mathematical and conceptual frameworks transform abstract principles into practical solutions across disciplines.

Exercise

1. Define Brownian motion and its mathematical properties. Provide a real-world example where Brownian motion plays a critical role.

2. What critical thinking ideas and elements of creative analogies stand out to you in this chapter?

3. Discuss the role of Brownian motion in adaptive clinical trial design. How does it help control false positive rates across multiple interim analyses?

4. Propose a scenario in oncology or immunology where threshold regression could be used to model time-to-event data. Explain how unobservable biomarkers and thresholds would be incorporated into the model.

5. Create an analogy to explain the connection between random walks and diffusion processes. Use a simple, everyday scenario to illustrate the relationship.

6. Develop an analogy for Monte Carlo simulations to explain their utility in solving complex problems. For example, liken it to repeatedly testing different routes to find the fastest way home.

7. Debate the advantages and limitations of using Brownian motion as a model in fields like finance or biology. Provide specific examples to support your arguments.

8. Explain how random walk methods can solve Laplace's equation. Provide a simple example, such as estimating the temperature distribution in a square domain.

9. Create network graph by connecting concepts discussed in this chapter through analogy and similarity: Stochastic process, adaptive clinical trial, discrete Brownian motion, random walk, microscopic level, threshold regression, Brownian motion, Wiener process, harmonic differential equation, quantum mechanics, Monte Carlo simulation, macroscopic level, mathematical model, Green's function, Lippmann–Schwinger equation, partial differential equation, Dirac delta function, Schrödinger equation, generalized function, continuous probability distribution, discrete probability distribution, and integral equation.

13

Quantum Computing Explained via Analogies

13.1 The Power of Analogies in Quantum Computing

Quantum computing encompasses two distinct but complementary approaches: quantum mechanics-based computing and quantum simulation.

13.1.1 Quantum Mechanics-Based Computing

1. This approach leverages quantum computers (hardware) to solve complex problems by exploiting quantum mechanical principles such as superposition, entanglement, and interference. Unlike classical computers, which use bits in definite states (0 or 1), quantum computers process information using qubits, which can exist in multiple states simultaneously. This enables significant computational speedups for tasks such as cryptography, optimization, and artificial intelligence.

2. Algorithms like Shor's algorithm (for integer factorization) and Grover's algorithm (for search optimization) exemplify the power of quantum hardware. These algorithms take advantage of entangled qubits to perform quantum parallelism and interference, making them crucial for tasks where quantum computers significantly outperform classical systems.

3. While classical computers can simulate quantum mechanics to a limited extent, such simulations become infeasible for larger systems due to exponential resource demands, underscoring the necessity of quantum hardware for realizing this approach's full potential.

13.1.2 Quantum Simulation

1. This approach focuses on modeling quantum systems to understand their behavior directly. Quantum simulations are widely used in chemistry, material science, and physics for tasks such as simulating

DOI: 10.1201/9781003630081-13

molecular interactions, studying quantum phase transitions, and designing new materials.

2. By replicating quantum phenomena, simulations provide insights into systems too complex for classical computation. Although computationally intensive, classical computers can perform small-scale simulations, such as solving the Schrödinger equation or simulating quantum circuits using specialized algorithms. However, for large systems, classical simulations face limitations due to the exponential growth in computational resource requirements.

3. Unlike quantum mechanics-based computing, which focuses on general-purpose problem-solving, quantum simulation specializes in advancing scientific discovery by targeting specific quantum systems.

Quantum computing is often perceived as an abstract and daunting field. Its foundation on counterintuitive principles like superposition, entanglement, and quantum interference challenges classical understanding. Bridging the gap between the unfamiliar quantum realm and our everyday experiences is essential—a task at which analogies excel.

At its core, quantum computing differs fundamentally from classical computing. Classical computers process information in binary units called bits, which exist in one of two states: 0 or 1. Quantum computers, on the other hand, use quantum bits or qubits, which can exist in multiple states simultaneously due to superposition. This property, along with entanglement and quantum interference, gives quantum computing its extraordinary power. Yet, these concepts defy everyday logic, requiring imaginative leaps to comprehend them.

For instance, superposition—the ability of a qubit to be in multiple states at once—can be compared to a spinning coin that embodies both heads and tails simultaneously, only settling on one when observed. Similarly, entanglement, where two qubits become deeply connected, can be likened to adjustable-length seesaws. Regardless of their physical separation, the movements of the two ends remain perfectly coordinated.

The value of analogies in quantum computing extends beyond education. Historically, they have driven major scientific advancements by offering new ways to visualize and approach problems. For example, Richard Feynman, one of the pioneers of quantum mechanics, famously used the analogy of a chess game to explain how nature's rules might differ at varying levels of observation. Similarly, a quantum gate, which manipulates qubits, can be likened to a verb in a sentence, transforming the "state" of the quantum language.

Analogical reasoning also reveals how quantum computing challenges classical intuitions. Classical computers are deterministic; their outcomes are predictable based on their inputs. In contrast, quantum systems thrive in uncertainty, probability, and coherence. Analogies serve as conceptual bridges, connecting the probabilistic nature of quantum mechanics to familiar uncertainties in decision-making.

The importance of analogical thinking extends to broader contexts of innovation and problem-solving. In the age of artificial intelligence and complex technologies, the ability to draw creative analogies is a critical skill. Analogies enable pattern recognition across disciplines, fostering interdisciplinary thinking essential for addressing the world's most challenging problems. Quantum computing, as one of the most transformative technologies of the 21st century, provides an ideal domain for cultivating such skills.

13.2 Superposition and Parallelism: The Symphony of Choices

Superposition is one of the most fascinating and challenging concepts in quantum computing. It represents a quantum system's ability to exist in multiple states simultaneously until it is observed or measured. This principle, which defies classical intuition, is key to the power of quantum computing. To understand superposition and its implications, analogies provide a valuable framework, offering a bridge between the quantum world and our everyday experiences.

At its simplest, superposition can be imagined as a spinning coin. While spinning, the coin embodies both heads and tails, existing in a state that is neither one nor the other, but potentially both. Only when the coin stops spinning and is observed does it "choose" a definite state. Similarly, a quantum bit, or qubit, in superposition exists in a blend of 0 and 1 states until it is measured. However, unlike the spinning coin, which is a probabilistic mix of two outcomes, a qubit's superposition is a precise mathematical combination described by a wave function, encompassing all possible outcomes with specific probabilities.

Superposition's implications for computing are profound. In classical computing, bits can represent either 0 or 1 at any given time. When performing calculations, classical computers must process each possible input sequentially or in parallel through additional hardware. In contrast, a quantum computer leverages superposition to process multiple inputs simultaneously within a single computational unit. This phenomenon, often called quantum parallelism, allows quantum computers to perform certain tasks exponentially faster than their classical counterparts.

To grasp the magnitude of this advantage, imagine a musician composing a symphony. A classical composer writes each note one by one, ensuring that every part of the symphony is crafted and played sequentially. A quantum composer, in contrast, can compose and play all possible variations of the symphony simultaneously, exploring and evaluating every combination in a single stroke. This ability to handle vast possibilities at once transforms problems that would take classical computers years to solve into tasks that quantum computers can complete in minutes.

Another analogy for superposition lies in decision-making. Consider a traveler standing at a crossroads, facing multiple paths. In the classical world, the traveler must choose one path and follow it, retracing their steps to try another route if necessary. A quantum traveler, however, can explore all paths simultaneously, gathering information from every route before deciding on the optimal journey. This idea underpins quantum computing's efficiency in solving problems like optimization, where multiple possibilities must be evaluated to find the best solution.

In practice, quantum algorithms harness superposition to tackle problems in fields as diverse as cryptography, logistics, and artificial intelligence. Shor's algorithm, for instance, uses superposition to factorize large numbers efficiently—a task critical for breaking classical encryption methods. Similarly, Grover's algorithm exploits superposition to search unsorted databases with unprecedented speed. These algorithms exemplify how quantum parallelism, enabled by superposition, can outperform classical approaches.

However, superposition is not without challenges. The act of measurement disrupts the superposition state, forcing the system to collapse into a single outcome. This phenomenon, known as wave function collapse, underscores the probabilistic nature of quantum mechanics. Designing quantum algorithms thus requires careful manipulation of superposition states to ensure that the desired outcome emerges with high probability upon measurement.

Superposition also invites us to rethink our understanding of reality. In the classical world, things exist in definite states; a light switch is either on or off. In the quantum world, reality is more fluid and probabilistic, challenging our intuitions about certainty and determinism. This shift in perspective opens new possibilities for thinking about problems, inspiring creative approaches in fields beyond computing.

In conclusion, superposition is the cornerstone of quantum computing's power, enabling quantum systems to process information in ways that defy classical constraints. By drawing on analogies such as spinning coins, musical symphonies, travelers at crossroads, and multidimensional projections, we can demystify this complex phenomenon and appreciate its revolutionary potential. Superposition not only reshapes our understanding of computation but also expands the horizons of human creativity and problem-solving.

13.3 Entanglement: The Quantum Tango

Entanglement is one of the most enigmatic and fascinating phenomena in quantum mechanics, often referred to as "spooky action at a distance" by Albert Einstein. It describes a situation where two or more particles become so deeply connected that the state of one instantly determines the state of the other, no matter how far apart they are. This peculiar behavior defies classical notions of separability and locality, making entanglement a cornerstone of quantum computing and a rich field for analogical exploration.

To understand entanglement, consider the analogy of an adjustable-length seesaw. Imagine a seesaw where the length of the plank can stretch infinitely, yet the balance between the two ends remains perfectly coordinated. If one person shifts their position or weight on one end, the other person's position is instantly adjusted to maintain the balance, no matter how far the seesaw stretches. This metaphor captures the essence of entanglement: a mysterious, intrinsic connection that transcends spatial separation.

Unlike classical systems, where information exchange requires physical proximity or communication signals, entangled particles seem to share information instantaneously. Using the adjustable-length seesaw as an analogy (Figure 13.1), the interaction between entangled particles reflects a dynamic

FIGURE 13.1
Adjustable-length seesaw as a metaphor for quantum entanglement.

relationship where changes on one end immediately and predictably influence the other. However, in the quantum realm, this interaction is more complex: entangled particles do not have fixed properties until measured. Their states remain undefined until observation, yet they are still perfectly correlated when observed, just like the seesaw maintaining balance regardless of its length.

This unique property of entanglement plays a crucial role in quantum computing. In classical computing, information is localized; each bit functions independently. Entanglement, however, allows qubits to share information in a deeply interconnected way. When qubits are entangled, the state of one qubit provides information about the states of others, enabling quantum computers to perform computations involving exponentially more data than classical systems.

Quantum entanglement also underpins revolutionary technologies like quantum cryptography and quantum teleportation. Quantum key distribution (QKD), for instance, uses entangled particles to create secure communication channels. The no-cloning principle, which states that quantum states cannot be perfectly copied, ensures that any attempt to intercept or duplicate the quantum key will disturb the entangled particles. This disturbance alerts the sender and receiver to the intrusion, making quantum communication fundamentally secure against eavesdropping—a significant advancement over classical encryption methods.

Quantum teleportation, another application of entanglement, involves transmitting quantum states from one location to another without physically moving the particles. This is achieved by using a pair of entangled particles and exchanging classical information about the state being teleported. While this technology does not yet allow for teleporting physical objects, it has profound implications for quantum networking and information transfer.

The implications of entanglement extend beyond computation and technology to the philosophical. Entanglement challenges our classical assumptions about the separateness of objects and the flow of time and causality. It invites us to rethink our understanding of connectedness, not just in physics but also in systems thinking, networks, and even human relationships.

13.4 Quantum Gates and Algorithms: The Language of Transformation

In classical computing, information is processed through logic gates, which manipulate binary bits to perform operations like AND, OR, and NOT. Similarly, in quantum computing, quantum gates act as the building blocks of computation, operating on qubits to transform their quantum states.

However, unlike classical gates, quantum gates utilize the principles of superposition and entanglement, enabling quantum computers to process information in ways that are fundamentally different—and far more powerful—than classical systems. To make sense of these gates and the algorithms they underpin, analogies can illuminate their abstract principles, much like a map guides a traveler through unfamiliar terrain.

13.4.1 Quantum Gates as the Verbs of Quantum Language

Imagine a sentence where words represent different components of thought. In this analogy, qubits are like nouns—they hold the "substance" of information, existing in states of 0, 1, or a superposition of both. Quantum gates, then, are the verbs: they act upon qubits, changing their states or relationships with one another.

For example, the NOT gate in classical computing flips a bit from 0 to 1 or 1 to 0. Its quantum counterpart, the X gate, similarly flips a qubit's state. However, quantum gates go far beyond binary flips. The Hadamard gate, for instance, transforms a qubit into a superposition state, akin to opening a door to infinite possibilities. Using our language analogy, this is like transforming a singular noun into a plural one, encompassing all possible variations of the original.

Another important gate, the CNOT gate (controlled-NOT) entangles two qubits in the process. It is as if one verb depends on the context of another noun, creating a relationship where changing one qubit's state automatically alters the other. These gates, and others like the phase gate and the Toffoli gate, form a quantum alphabet capable of expressing complex transformations.

13.4.2 Algorithms as Orchestrated Transformations

Just as words combine to form sentences, quantum gates combine to create quantum circuits, which implement quantum algorithms. These algorithms are sequences of gate operations designed to solve specific problems. Unlike classical algorithms, which often rely on step-by-step procedures, quantum algorithms leverage parallelism and interference to achieve results more efficiently.

One of the most famous quantum algorithms is Shor's algorithm, which factors large numbers exponentially faster than classical methods. This capability has profound implications for cryptography, as it can break encryption schemes based on the difficulty of factorization. Analogically, Shor's algorithm is like a master locksmith who, instead of trying every possible key sequentially, uses a magical tool to simultaneously test all keys and instantly find the right one.

Another key algorithm is Grover's algorithm, which excels at searching unsorted databases. Imagine you are searching for a name in a phone book with no order. A classical computer checks each entry one by one, but Grover's algorithm acts like a spotlight, illuminating the desired name in far fewer steps. This remarkable efficiency stems from quantum interference, where the algorithm amplifies correct solutions while canceling out incorrect ones.

13.5 Quantum Measurement: The Observer's Dilemma

At the heart of quantum mechanics lies a profound mystery: the role of measurement. Quantum systems, unlike their classical counterparts, exist in superpositions of states—a blend of possibilities described by a wave function. However, the act of measurement forces the system to "collapse" into one specific state. This phenomenon raises a fundamental question: what role does the observer play in determining reality? Known as the observer's dilemma, this concept not only defines quantum mechanics but also challenges our understanding of objectivity, causality, and determinism. Analogies can help us unpack this complex interplay between observation and reality.

Perhaps the most famous analogy for quantum measurement is Schrödinger's cat, a thought experiment devised by physicist Erwin Schrödinger to illustrate the perplexities of quantum superposition and measurement. Imagine a cat placed inside a sealed box with a radioactive atom, a Geiger counter and a vial of poison. If the atom decays, the Geiger counter triggers the release of poison, killing the cat. If the atom does not decay, the cat remains alive. Quantum mechanics suggests that, until the box is opened and the cat observed, it exists in a superposition of being both alive and dead.

The act of opening the box constitutes a measurement, collapsing the superposition into one of two outcomes: alive or dead. This paradox highlights the peculiar role of the observer in determining the state of a quantum system. Unlike classical systems, where properties exist independently of observation, quantum systems seem to depend on the act of measurement for their definitive states.

13.5.1 Quantum Measurement and the Loss of Information

Another useful analogy is flipping a coin and catching it in your hand without looking at it. Until you observe the coin, its outcome remains undetermined—it could be heads or tails. However, this classical analogy falls short in capturing the deeper implications of quantum measurement. In the quantum world, the "coin" doesn't have a definitive side while unobserved; it exists in a superposition of heads and tails. Measurement does not simply reveal a pre-existing state; it fundamentally alters the system, forcing it into a specific state and erasing the possibility of the other outcomes.

This loss of information due to measurement is a defining feature of quantum mechanics. Once the wave function collapses, the rich tapestry of probabilities encoded in the superposition is reduced to a single reality. This irreversibility of measurement distinguishes quantum mechanics from classical systems and underscores the delicacy of quantum information.

13.5.2 The Observer's Influence: Active or Passive?

The idea that observation collapses the wave function raises an intriguing philosophical question: does the observer actively influence the system, or is measurement a passive process that merely registers an outcome? In classical physics, measurement is a neutral act—reading the temperature of a cup of coffee does not change the coffee's heat. In quantum mechanics, however, the act of measuring fundamentally changes the system being measured.

The question of whether an observer can influence the outcome of a quantum system hinges on how we define "sameness." If sameness is defined rigidly—requiring a quantum entity to maintain identical states across observations—the observer's influence appears negligible, as the outcome is determined solely by the system's inherent properties. However, if sameness is context-dependent, allowing the quantum entity to belong to the same type but vary based on interaction, then the observer influences the outcome by shaping the measurement context. This distinction highlights that the observer's role is not universally active or passive but relational, dependent on philosophical interpretations of identity. Thus, the observer's influence becomes a matter of how identity is framed within the quantum system.

13.5.3 Implications for Quantum Computing

Quantum measurement poses unique challenges and opportunities for quantum computing. The power of quantum algorithms lies in their ability to process information in superposition. However, to extract useful results, measurement is inevitable, collapsing the system and reducing the superposition to a single state. The key to successful quantum computation is designing algorithms that maximize the probability of the desired outcome before measurement.

When the system is measured, the desired result emerges with high probability. This delicate balance between computation and measurement exemplifies the importance of understanding and manipulating the observer's dilemma in practical applications.

13.5.4 Challenges and Future Directions

Quantum measurement, with its observer's dilemma, is both a challenge and a gateway to understanding the quantum world. While the power of quantum gates and algorithms is immense, implementing them on a large scale presents significant challenges. Quantum systems are highly sensitive to environmental interactions, which can lead to errors. Advances in quantum error correction and fault-tolerant computation are essential to unlocking the full potential of quantum computing. Despite these hurdles, the development of quantum gates and algorithms continues to push the boundaries of what is computationally possible.

13.6 The Four-Dimensional World Theory

Quantum mechanics reveals several extraordinary features that underpin quantum computing, each of which highlights the limitations of classical perspectives and the possibilities of higher-dimensional interpretations:

13.6.1 Key Quantum Features in Quantum Computing

- Superposition: Enables qubits to exist in multiple states simultaneously, allowing parallel computation and exponential speedups for certain problems.
- Entanglement: Creates correlations between qubits, enabling complex, interconnected computations and enhancing efficiency in algorithms like quantum cryptography.
- Quantum interference: Amplifies correct solutions and cancels out incorrect ones in algorithms, improving problem-solving accuracy.
- Quantum tunneling: Aids optimization problems by allowing systems to bypass energy barriers, exploring solutions more efficiently.
- Quantum parallelism: Processes many possibilities at once, drastically reducing the time for specific computations compared to classical methods.
- Quantum algorithm: Just as words combine to form sentences, quantum gates combine to create quantum circuits, which implement quantum algorithms to solve specific problems.

These features, while profound, challenge classical intuitions. To better interpret them, I propose the 4D world theory, a conceptual framework that uses higher-dimensional analogies to explain quantum phenomena like superposition and entanglement.

13.6.2 The Sensor-Dependent Multifaceted World

Our understanding of the world is intricately tied to the capabilities of our sensory organs. Each being perceives reality within the constraints of its

a) A cartoon metaphor illustrating a summer mosquito enjoying its brief life in warmth, surrounded by a sunny meadow. In the distance, an unreachable winter landscape symbolizes its inability to comprehend the colder season.

b) A cartoon metaphor of an adult man thoughtfully trying to 'hear' music by interacting with a sleek speaker emitting vibrant soundwaves and tactile vibrations.

c) A metaphorical illustration of a blind person in a colorful world, emphasizing their deep connection to the environment through other senses.

FIGURE 13.2

Sensory organs shaping perceived worlds: a metaphorical view.

sensory limitations, leaving much of the universe unseen, unheard, or unfelt. This sensor-dependent nature creates a multifaceted world where each perspective is uniquely limited and uniquely valid.

Consider a summer mosquito. It lives its brief life in the warmth of the season, entirely incapable of comprehending what winter might feel like. A mosquito might only indirectly understand winter if a higher being—like a god—were to describe it. Similarly, a one-season creature cannot grasp the cyclic beauty of a world that spans all four seasons. Its sensory reality confines it to a singular facet of the multifaceted world (Figure 13.2a).

A deaf individual may imagine music but cannot directly experience its rhythms or melodies. They might construct a mental picture of sound from descriptions or vibrations, but the rich tapestry of tones and harmonies lies beyond their sensory grasp (Figure 13.2b). Imagine, then, a human species evolved entirely from color-blind ancestors. Their perception of the world would lack the vibrancy of color. How profoundly would their understanding of art, nature, and even emotion differ from ours (Figure 13.2c)?

This concept of sensory limitations is eloquently explored in Edwin A. Abbott's *Flatland: A Romance of Many Dimensions*. In Flatland, two-dimensional beings live in a plane where they perceive only length and width, utterly unaware of the existence of height. Their reality is bound to their plane, and the idea of a three-dimensional world is incomprehensible to them until a higher-dimensional being introduces it.

Now imagine humans evolving further, gaining sensory organs capable of perceiving beyond three dimensions. Such superbeings could experience a 4D world, perceiving facets of reality entirely hidden from our current senses.

The drawing style of this illustration is **minimalist cartoon art** with a **playful and educational tone**. It uses simple shapes, smooth shading, and expressive characters to convey abstract concepts like dimensionality. The clean lines, vibrant colors, and exaggerated features (like large eyes) make it approachable and engaging, perfect for visualizing complex ideas in a fun and accessible way.

FIGURE 13.3
The creature imagines itself in a higher-dimensional world.

Just as color adds richness to vision, higher dimensions could reveal complexities of time, space, and existence that are currently beyond our imagination.

The multifaceted world we perceive is only a fraction of what might exist. Our sensory limitations define our reality, but they also inspire us to imagine what lies beyond—pushing the boundaries of knowledge and understanding. Whether through evolution, technological augmentation, or intellectual leaps, the journey to experience higher dimensions is a profound reflection of humanity's innate curiosity (Figure 13.3).

The 4D world theory retains the fundamental principles of quantum mechanics, such as Schrödinger's wave function and the Lippmann–Schwinger equation, but offers a more intuitive and supportive interpretation compared to Schrödinger's duality cat or the multiple-world theory. By providing a higher-dimensional perspective, it bridges abstract quantum phenomena with more tangible, multidimensional analogies, enriching our understanding of quantum reality.

13.6.3 Superposition and the 4D Cat Analogy

Imagine a 4D world where a sequence of cats exists, each with distinct colors. When projected into our 3D world, this sequence appears as a single cat in superposition, embodying all colors probabilistically. Just as a 2D being might struggle to comprehend how a 3D object enters a ring without breaking it, our limited 3D perception interprets the blended projection

FIGURE 13.4
Cats passing through a portal in a higher-dimensional world.

as a superposition. Measurement in this analogy corresponds to observing the projection at a specific point, determining the cat's visible color. In this 4D analogy, wave function collapse is akin to fixing the projection of the sequence at a single point, where the observer perceives one specific outcome out of the blended possibilities (Figure 13.4).

Humans in 3D cannot fully grasp the 4D dynamics shaping quantum systems, but by leveraging the fourth dimension, a sequence of color cats' explanation of superposition becomes much easier to understand.

Measurement in this analogy corresponds to observing the projection of the sequence at a specific point in 3D space, determining the cat's visible color. Wave function collapse, then, is akin to fixing the projection at a single point, where one specific outcome is perceived out of the blended possibilities. This analogy offers a compelling way to conceptualize the probabilistic nature of quantum measurement and the relationship between dimensionality and observation.

13.6.4 Quantum Entanglement and Proximity in 4D Space

In the 4D world, two particles that appear distant in the 3D world could be very close or directly connected. This perspective offers an intriguing explanation for quantum entanglement, where the instantaneous correlation between particles—known as "spooky action at a distance"—may arise from their proximity or unified structure in higher-dimensional space.

1. **Spatial Proximity in Higher Dimensions**
   - **Higher-dimensional shortcut**: In 4D space, two entangled particles might occupy points that are close to each other or directly connected via the additional dimension. In 3D, this connection appears as an instantaneous correlation over large distances.
   - **Analogy**: Consider two dots on opposite sides of a flat sheet (2D). In the 3D world, the sheet can be folded so the dots are adjacent. Similarly, in 4D, the apparent distance between entangled particles in 3D vanishes.

2. **Implications for Entanglement**
   - **Nonlocality explained**: The instantaneous correlation observed in entangled particles could result from their intrinsic proximity in 4D space, eliminating the need for faster-than-light communication.
   - **Unified wavefunction**: In 4D, entangled particles may form a single, unified quantum entity. Measurement of one particle corresponds to observing a specific part of this entity, instantly defining the state of the other.

13.6.5 Broader Implications of the 4D World Theory

1. **Quantum correlations:** The higher-dimensional perspective explains why quantum correlations persist regardless of 3D distance. Particles that seem "far apart" in 3D are not truly distant in their 4D configuration.

2. **Efficient information transfer:** Information about one particle's state does not need to travel across space; it already exists within the shared higher-dimensional state. This makes the entanglement "instantaneous" from a 3D perspective.

3. **Experimental predictions:** If entangled particles are close in 4D, experimental setups probing additional dimensions (e.g., quantum field theories or brane cosmology) might reveal patterns suggesting higher-dimensional proximity. Testing these effects could provide deeper insights into the nature of quantum entanglement.

4. **Potential analogies for exploration:** Models of entangled particles can simulate the effects of proximity in higher dimensions through dimensional folding or projection, offering new ways to visualize and study quantum systems.

In conclusion, the 4D world theory offers a fresh perspective on quantum phenomena, using higher-dimensional analogies to bridge the gap between classical intuition and quantum reality. By conceptualizing superposition and entanglement as projections of higher-dimensional states into our 3D world, it sheds light on the probabilistic and interconnected nature of quantum systems. This approach not only enhances our understanding of quantum mechanics but also opens the door to innovative interpretations and experimental exploration.

13.7 Future Analogies: Quantum Thinking for Problem Solving

13.7.1 Quantum Interference: Constructive and Destructive Problem Solving

Quantum interference, the process by which probabilities are amplified or canceled out through the interplay of wave functions, can inspire strategies for refining solutions. In problem-solving, interference can be viewed as the interplay of ideas, where some concepts reinforce and others eliminate each other.

For instance, in team settings, constructive interference occurs when diverse perspectives build upon one another to create innovative solutions. Destructive interference, on the other hand, eliminates redundant or counterproductive ideas. Applying this principle, organizations can design workflows and decision-making processes that foster constructive collaboration while filtering out inefficiencies, mirroring the precision of quantum algorithms.

13.7.2 Quantum Algorithms as Frameworks for Creativity

Quantum algorithms such as Grover's and Shor's offer direct inspiration for structured approaches to problem-solving. Grover's algorithm, which efficiently searches unsorted databases, can be analogized to methods for prioritizing resources in a chaotic environment. Similarly, Shor's algorithm, which factors large numbers, can inspire techniques for breaking down seemingly insurmountable challenges into manageable components.

These algorithms exemplify the importance of leveraging unique tools for specific tasks. A quantum-inspired thinker would analyze problems not through brute force but by identifying patterns, exploiting symmetries, and designing tailored strategies—an approach valuable in fields ranging from artificial intelligence to economics.

13.7.3 A Quantum-Inspired Future

The true potential of quantum thinking lies in its ability to reshape our problem-solving paradigms. As quantum computing evolves, it will drive a shift toward interdisciplinary approaches that combine physics, computer science, mathematics, and philosophy. Analogies will continue to play a central role in this evolution, helping us translate quantum principles into actionable strategies for the real world.

13.8 Summary: Quantum Computing and Analogical Exploration

This chapter delves into the transformative potential of quantum computing, using creative analogies to make its abstract principles—such as superposition, entanglement, and quantum gates—accessible and intuitive. By bridging quantum mechanics with everyday experiences, the chapter explores two complementary approaches: quantum mechanics-based computing, which solves complex problems using algorithms like Shor's and Grover's, and quantum simulation, which models quantum systems for scientific discovery. The chapter also introduces the 4D world theory, a novel framework for interpreting quantum phenomena through higher-dimensional analogies.

Key Takeaways from Critical Thinking Creative Analogy Perspectives

1. Interdisciplinary thinking: The chapter advocates for combining physics, computer science, and philosophy to reimagine problem-solving frameworks, demonstrating how quantum principles inspire innovation across fields.

2. Entanglement: Likened to an adjustable-length seesaw, highlighting the intrinsic connection between particles regardless of distance.

3. Multidimensional analogies: Introduces higher-dimensional interpretations to explain quantum phenomena, such as superposition as a projection of 4D states into 3D space. Use cats passing through a portal to visualize quantum measurement and wave function collapse using a higher-dimensional metaphor.

4. Algorithmic insights: By exploring quantum algorithms, the chapter showcases how quantum principles like interference and parallelism can optimize problem-solving, offering lessons for creativity and structured thinking.

5. Quantum algorithms: Grover's algorithm is analogized to a spotlight illuminating solutions, while Shor's algorithm is likened to a master locksmith testing all keys simultaneously, underscoring their efficiency.

6. Efficiency vs correctness: Quantum computing balances computational efficiency and probability of correctness. Algorithms can often run faster than classical counterparts, but they may need repeated executions, post-processing, or error correction, which adds computational overhead.

This chapter integrates critical thinking and creative analogies, providing a foundation for understanding quantum computing while inspiring innovative approaches to complex problems in science, technology, and philosophy.

Exercise

1. Define superposition and entanglement in the context of quantum computing. Use examples or analogies from the chapter to illustrate these concepts.

2. Compare and contrast classical and quantum computing in terms of their basic units (bits vs. qubits), processing methods, and computational potential.

3. What critical thinking ideas and elements of creative analogies stand out to you in this chapter?

4. Explain the spinning coin analogy for superposition and the adjustable-length seesaw analogy for entanglement. How do these analogies help bridge classical and quantum intuitions?

5. Describe how Grover's algorithm and Shor's algorithm utilize quantum principles to outperform classical approaches. Provide a real-world application for each.

6. Use the 4D world theory to explain quantum superposition, observer's dilemma, and entanglement. Create a unique analogy that extends the concepts provided in the chapter.

7. Develop a new analogy to explain constructive and destructive quantum interference. How could this analogy be applied to team problem-solving or decision-making processes?

8. Debate whether quantum measurement is an active or passive process. How does this affect our understanding of reality and the observer's role in quantum mechanics?

9. Discuss the philosophical implications of quantum phenomena, such as superposition and entanglement, on our understanding of reality and causality.

10. Design a simple quantum circuit using gates like Hadamard and CNOT. Explain the function of each gate and the transformation it applies to the qubits.

11. Imagine a practical application of quantum-inspired problem-solving in a non-technical field, such as healthcare or education. How could principles like quantum parallelism or interference revolutionize the field?

14

Dual Forces: Evolution, Devolution, and Feedback Systems

The evolution and devolution duality underscores the simultaneous progression and regression within interconnected entities, illustrating life's inherent paradox.

14.1 Evolution by Natural Selection

14.1.1 Understanding the Concept

Evolution discussed in the literature has focused almost exclusively on competition among separate, parallel entities. However, if we consider competition between an entity and its own parts, devolution and evolution can be equally likely.

The three fundamental conditions for evolution by natural selection are as follows:

1. Variation—Individuals within a population differ in traits.
2. Heritability—These differences must be genetically passed to offspring.
3. Differential reproductive success—Individuals with advantageous traits leave more offspring than others.

Limited resources create competition, which drives differential reproductive success. It ensures that not all individuals survive and reproduce equally, making certain traits more favorable over time.

Thus, limited resources are a selective pressure, not a separate condition. Evolution can still occur in situations where competition is indirect (e.g., predator avoidance and mate selection) without resource scarcity.

There is no reason to believe that devolution does not occur within or between organisms. In fact, it can happen at different levels for various reasons:

1. Medicine can cure diseases and extend life for weaker individuals, but it may simultaneously weaken the human immune system.

DOI: 10.1201/9781003630081-14

2. As environments change, an organism well-suited to one may not thrive in another. For example, cyclical changes (like the four seasons) can cause the "fittest" organisms to shift with the environment.

3. Evolution can occur at a higher societal level, while devolution may happen at lower levels, such as within individuals—and vice versa.

Since "fitness" is multifaceted, it is often subjective and difficult to define. Consider two couples: one is healthy and expected to live long but has low fertility, while the other is less healthy but highly fertile and chooses to have more children. Which couple will "win" through evolution over generations? Critics of Darwin's "survival of the fittest" argue that the concept is tautological—whatever survives is deemed "fit." Supporters, however, contend that fitness can be empirically measured by traits like speed, strength, disease resistance, and aerodynamic efficiency (for flying species), independent of mere survival (Gintis, 2009).

14.1.2 Controversies of Fitting

Andy, Bob, and Charlie engage in a shootout. Each player takes turns shooting until only one remains. The shooter can target any of the other players or shoot into the air. The order is decided by drawing straws at the start of the game, and they continue in that sequence. The goal for each player is to survive as long as possible. Andy has 50% accuracy, Bob 80%, and Charlie 100%.

There are two scenarios:

1. Unknown skills: Players are unaware of others' shooting skills and shoot randomly, aiming to hit a person each time.

2. Known skills: Players are fully aware of each other's abilities and always target the most dangerous player.

In Scenario 1 (random targeting), the survival probabilities for Andy, Bob, and Charlie are 21.7%, 34.3%, and 44%, respectively. However, in Scenario 2, the survival chances are surprisingly inverse to their accuracy: Andy survives 45% of the time, Bob 31%, and Charlie 24%. Remarkably, Bob and Charlie fare worse than in random shooting and would be better off cooperating (powerful rivals form alliances) to eliminate Andy first (Figure 14.1).

This strategic setup, where the weakest player (Andy) survives the most often because the stronger players target each other first, mirrors several real-life scenarios where the weaker party benefits from the competition or conflict between stronger rivals. Here are a few analogies that highlight how similar dynamics play out in real life:

1. Corporate Rivalry and Startups
   - Analogy: Imagine Andy as a small startup with limited market share, while Bob and Charlie represent two tech giants with

The Least Accurate Shooter Is Most Likely to Survive the Longest.

FIGURE 14.1
Survival paradox: The weakest lives the longest.

dominant market positions. Bob and Charlie (the tech giants) see each other as direct threats due to their large customer bases and cutting-edge products, so they focus on competing with each other to maintain market dominance. This rivalry leaves Andy (the startup) relatively ignored and free to grow in niche areas, picking up customers slowly but surely while the giants are busy.

- Real-life example: Companies like Netflix and Uber initially operated in niche markets without significant threats from established players. While competitors in mainstream media and transportation targeted each other, these smaller players capitalized on the opportunity to build user bases and establish themselves.

2. Political Strategy in Elections

- Analogy: In a three-way election with two strong candidates (Bob and Charlie) and one weak candidate (Andy), the stronger candidates often focus on each other, leaving the weaker candidate to gain ground unnoticed. The two frontrunners might attack each other's policies and character, while the weaker candidate can avoid criticism and appeal to a different voter base, potentially capturing undecided voters.

- **Real-life example**: In many political systems, smaller third-party candidates can grow in popularity when the major candidates focus their resources on each other. For example, in U.S. elections, third-party candidates have occasionally gained traction by positioning themselves as an alternative while major parties target each other directly.

3. Cold War Dynamics and Proxy Conflicts

- **Analogy**: During the Cold War, the United States (Charlie) and the Soviet Union (Bob) were in constant competition for global influence, often viewing each other as the main threat. Smaller countries (like Andy) sometimes benefited by remaining neutral or by playing both sides to gain economic or military aid from the superpowers, who were more focused on each other.

- **Real-life example**: Countries like Egypt and Yugoslavia were able to gain significant aid from both the U.S. and the Soviet Union during the Cold War, as the superpowers sought to increase influence in these regions while primarily focusing on countering each other's actions.

In each analogy, the weaker player benefits by staying out of the spotlight or letting the stronger players target each other. This counterintuitive strategy, where strength leads to greater vulnerability, is common in competitive environments and situations involving conflict.

Individual self-interest and micro-level motivations can sometimes create a devolutionary force in society. The weaker—or more cautious—individuals may find ways to survive longer, often at the expense of overall progress or strength. This could lead to a society where strength is no longer the primary determinant of success, a situation that seems to counteract the traditional evolutionary model of "survival of the fittest."

However, this presents an ethical dilemma: Should every member of society have an equal chance at survival, regardless of strength or ability? While equity may seem at odds with evolutionary principles, it could also serve as a key measure of a "better society." In the long run, a truly evolved society may find a balance between rewarding the strong and safeguarding the vulnerable.

14.2 Weaker Individuals Make a Stronger Team

A child wants to earn a spot on the local baseball team by proving they can get two consecutive hits during batting practice. Two people will pitch to the child:

- Parent A (the stronger pitcher), against whom the child's probability of hitting the ball is p.

- Parent B (the weaker pitcher), against whom the probability is q.
- We know $p<q$, since Parent A is more difficult to hit.

The child is allowed to choose between two sequences:

1. A–B–A (Parent A, then B, then A again)
2. B–A–B (Parent B, then A, then B again)

Intuition suggests choosing B–A–B, because the child faces the tougher pitcher (A) only once in the middle. However, a simple probability calculation shows that A–B–A is actually more favorable. Here is why:
For Sequence A–B–A,

- The child can succeed by hitting (A, B) or (B, A).
- Probability of hitting (A, B) in consecutive pitches: pq.
- Probability of hitting (B, A) in consecutive pitches: qp.
- Since the middle pitch is B, there's also a slight overlap if the child happens to succeed in all three, so we subtract the triple overlap pqp.
- Total winning probability: $pq+qp-pqp$.

Similarly, for Sequence B–A–B,

- Total winning probability: $qp+pq-qpq$.

The difference between the two winning probabilities is $qpq-pqp>0$, reaching its maximum of 0.25 when $p=0.5$ and $q=1$. Consequently, the A–B–A sequence yields a higher chance of achieving two consecutive hits. Counterintuitively, having more "weaker" matchups can produce a stronger overall outcome, while having fewer "weaker" matchups—that is, facing stronger opponents less often—can paradoxically weaken one's chances.

14.3 Horse Race Game and Strategies of Survival

Strategies of game play are important to winning or survival as the examples illustrated below.

14.3.1 Game Setup

- Two teams (Team A and Team B) each have a similar set of three horses ranked by speed: fast (F), medium (M), and slow (S).

- Teams assign their horses to three one-on-one races.
- The team that wins at least two races wins the match.

14.3.2 Strategies

1. Naïve strategy (matching strengths)
   - Each team assigns fast vs. fast, medium vs. medium, slow vs. slow.
   - Often results in a 1–1 tie before the final race, making it risky.
2. Optimized strategy (strategic sacrifice)
   - Sacrifice the weakest horse by racing it against the opponent's fastest horse.
   - Ensure favorable matchups in the remaining two races: fastest vs. medium vs. slowest.
   - This increases the probability of securing a 2–1 victory.

The game theory insights of the Chinese Horse Race problem reveal several strategic principles. The minimax strategy minimizes the risk of relying on a final-race decider by ensuring a more controlled path to victory. This strategic framework has real-world applications in various fields, including sports, business, negotiations, military tactics, and investment decision-making. Without knowing the sequence of horses the other player places, the only Nash equilibrium is a mixed strategy equilibrium where each team chooses their sequence uniformly at random.

14.4 Hierarchical Evolution-Devolution

Multiple-level evolution paradox: Evolution and devolution are intertwined at different levels within the same entity. If evolution involves removing unfavorable traits across generations, it inherently leads to the degeneration of certain sub-entities. For example, the prolonged survival of a cancer cell (or oncogenes) may result in a shorter lifespan for the individual hosting it (Figure 14.2). This duality highlights the tension between the survival of individual components and the overall health of a system, whether an organism or society, suggesting that devolution and evolution are equally probable in principle.

The cancer cell example highlights how evolution and devolution can indeed occur at any level, depending on how we subjectively define the entity and its subsystems. Evolution often implies progress or adaptation at one level (e.g., cancer cells evolving to survive and proliferate), while devolution

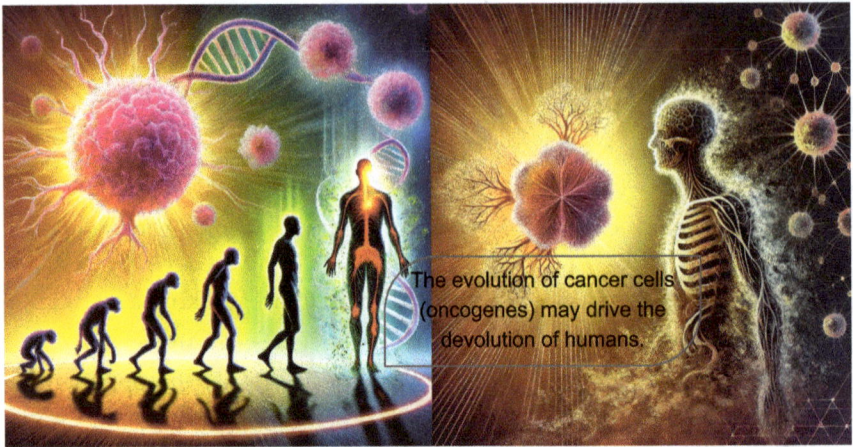

FIGURE 14.2
Hierarchical evolution and devolution: a metaphorical representation.

occurs simultaneously at another level (e.g., the overall decline in the health of the organism).

Since entities—whether they are individuals, cells, or social systems—can be defined at different scales, evolution and devolution are not confined to a single level. The processes that promote survival or advancement in one context may lead to decline or failure in another. This duality suggests that the interplay of evolution and devolution is relative, context-dependent, and intertwined across various levels of complexity, whether biological, social, or even psychological. In general, if a "diseased" part survives longer, the entity tends to die sooner.

Devolution and evolution can also arise from human-driven processes, even those grounded in scientific or statistical methodologies. Consider the example of drug evaluation and devolution. As new drugs are continually developed, it is reasonable to expect that treatments for a given disease will progressively improve over time. However, the reality is more nuanced. Rarely but inevitably, during the drug development process, an ineffective drug may be falsely deemed superior to the existing standard treatment—the best drug on the market—due to statistical or methodological errors.

When this occurs, the inferior drug may erroneously become the new standard or "best" treatment for the disease, resulting in devolution rather than progress. This phenomenon can arise even under rigorous statistical controls, such as maintaining a 2.5% false positive error rate. Consequently, evolution (more likely) and devolution (less likely) can alternate during the drug development process, underscoring the challenges in ensuring consistent progress in medical treatments.

In essence, we can think of evolution and devolution as **coexisting forces** that shape outcomes depending on how we frame the system. Thus, the boundary between evolution and devolution is fluid, and these processes can operate simultaneously at different levels within any given entity.

14.5 Duality of Feedback Systems

Feedback mechanisms play a crucial role in system survival. Positive (negative) feedback systems feature a positive (negative) feedback mechanism. Both systems represent fundamental mechanisms of regulation in natural, technological, and social contexts. Positive feedback amplifies deviations, often leading to exponential changes or extremes, while negative feedback counteracts deviations to restore stability and balance.

For instance, childbirth's hormonal loops exemplify positive feedback's reinforcing nature. Conversely, thermostats and biological processes like blood glucose regulation embody negative feedback's stabilizing role. By leveraging analogy, we can better grasp how these mechanisms operate and generalize their principles to other domains, whether explaining environmental tipping points or designing resilient systems. Understanding the interplay of positive and negative feedback through analogy not only enhances comprehension but also highlights their broader implications in science, engineering, and policymaking.

14.5.1 Positive Feedback Systems

Positive feedback systems amplify deviations from a setpoint, reinforcing the original stimulus rather than counteracting it. Positive feedback tends to drive systems toward extremes, often resulting in rapid changes or "runaway" effects. Desirable positive feedback systems result in beneficial or necessary outcomes, often with built-in mechanisms for termination.

In biology, positive feedback mechanisms enhance processes critical for survival and reproduction. For instance, during childbirth, uterine contractions stimulate the release of oxytocin, which intensifies contractions further. This feedback loop continues until the baby is delivered, after which it self-terminates. Similarly, blood clotting involves platelets attracting more platelets to a wound site, forming a clot that seals the injury. Once the wound is closed, this feedback mechanism stops. Another example is the immune response, where B-cells encountering an antigen proliferate rapidly, generating a large number of antibodies. This amplification ensures a robust response to infection.

In technological adoption, network effects create positive feedback loops. As more people adopt a technology, its value increases, attracting even more users. This phenomenon is evident in the growth of social media platforms. In education, positive feedback occurs when initial success boosts a student's confidence and motivation, leading to further effort and even greater success. This creates an upward spiral of engagement and achievement.

Undesirable positive feedback systems, on the other hand, can lead to instability, damage, or negative consequences.

In financial markets, positive feedback can drive speculative bubbles. Rising asset prices attract more investors, driving prices even higher until the bubble bursts, often resulting in a market crash. Similarly, hyperinflation occurs when rising prices lead to increased wage demands, which in turn drive prices even higher, creating a runaway inflationary spiral.

Social dynamics also demonstrate undesirable positive feedback. For instance, social polarization is reinforced by echo chambers on social media, which amplify existing beliefs, leading to increased division and societal fragmentation. Escalating conflicts provide another example: a small dispute rapidly grows as each side retaliates, potentially culminating in full-scale war.

In psychology, positive feedback loops can manifest in cycles of anxiety or panic. For example, a person experiencing anxiety might interpret physical symptoms, such as a racing heart, as evidence of a serious problem. This misinterpretation intensifies their anxiety, which in turn exacerbates the physical symptoms, creating a vicious cycle that spirals out of control. Left unchecked, this feedback loop can lead to full-blown panic attacks or chronic anxiety disorders.

To summarize, positive feedback systems, while powerful, require careful consideration due to their potential for both beneficial and harmful effects. Understanding their mechanisms is crucial for leveraging their advantages and mitigating their risks.

14.5.2 Negative Feedback Systems

14.5.2.1 Negative Feedback Systems in Biology and Physiology

Negative feedback systems are prevalent in biology and human physiology. These systems are crucial for maintaining homeostasis—stable internal conditions essential for survival and optimal function.

Thermoregulation involves the body maintaining a stable internal temperature of around 37°C (98.6°F) through negative feedback. When the temperature rises, such as during exercise or exposure to heat, the hypothalamus activates mechanisms like sweating and vasodilation to dissipate heat. Conversely, if the temperature drops, the hypothalamus triggers shivering and vasoconstriction to conserve and generate heat. This feedback loop ensures that the body remains within a narrow temperature range, crucial for enzymatic and metabolic processes to function effectively.

Blood glucose regulation is managed by the pancreas through negative feedback to maintain stable glucose levels. After eating, high blood sugar prompts the pancreas to release insulin, which facilitates glucose uptake by cells and storage as glycogen in the liver, reducing blood sugar. During fasting or physical activity, low blood sugar triggers the pancreas to release glucagon, stimulating glycogen breakdown into glucose in the liver, raising blood sugar levels. This system prevents harmful conditions like hyperglycemia and hypoglycemia.

Hormonal regulation, such as the hypothalamic–pituitary–thyroid axis, operates via negative feedback. The hypothalamus releases thyrotropin-releasing hormone (TRH), which stimulates the pituitary gland to secrete thyroid-stimulating hormone (TSH). TSH then prompts the thyroid gland to produce thyroid hormones (T3 and T4). Elevated levels of T3 and T4 signal the hypothalamus and pituitary to reduce TRH and TSH production, maintaining hormonal balance essential for metabolism and growth.

Blood pressure regulation is governed by the baroreflex, where baroreceptors in the carotid arteries and aorta monitor blood pressure. High blood pressure activates signals to the brain, leading to vasodilation and a reduced heart rate to lower pressure. Conversely, low blood pressure triggers vasoconstriction and an increased heart rate to raise it. This feedback loop protects blood vessels from hypertension damage and ensures adequate flow during hypotension.

Oxygen and carbon dioxide levels are regulated through respiratory feedback. High CO_2 levels in the blood are detected by chemoreceptors, prompting the brain to increase the breathing rate and depth to expel excess CO_2. If CO_2 levels are too low, such as after hyperventilation, breathing slows to restore balance. This system ensures proper gas exchange in the lungs and maintains blood pH and oxygen supply.

Calcium regulation is managed by the parathyroid gland. When calcium levels are low, the gland releases parathyroid hormone (PTH), which stimulates calcium release from bones, absorption in the intestines, and reabsorption in the kidneys. When calcium levels are high, PTH secretion decreases. This balance is vital for nerve function, muscle contraction, and bone health.

Emotional and behavioral regulation in humans involves negative feedback in the brain. For example, during stress, the hypothalamus triggers cortisol release. Elevated cortisol levels eventually signal the hypothalamus and pituitary to reduce its release, stabilizing the response. Cognitive processes also help regulate emotional responses to stimuli, preventing overreaction and maintaining mental and physical health.

Hunger and satiety are regulated by hormones like ghrelin and leptin. High hunger levels stimulate ghrelin release from the stomach, signaling the brain to increase appetite. After eating, fat cells release leptin, signaling the brain to reduce appetite. This system balances energy intake with needs, preventing overeating or starvation.

pH balance in the blood is maintained at approximately 7.4 through the bicarbonate buffer system. When blood pH drops (acidosis), the respiratory system increases breathing to expel CO_2, raising pH. Conversely, when pH rises (alkalosis), breathing slows, retaining CO_2 and lowering pH. This system ensures proper enzyme activity and cellular function.

In conclusion, negative feedback systems in biology are essential for stability and adaptation, enabling precise regulation of physiological processes necessary for survival and optimal function. Disruptions or malfunctions in these systems, such as insulin resistance in diabetes, can lead to significant health issues, highlighting their critical role in maintaining homeostasis.

14.5.2.2 Feedback Systems in Technology

Maxwell's mathematical analysis of control systems in 1868 laid the theoretical foundation for understanding feedback in mechanical systems. Norbert Wiener expanded the concept in his 1948 work on cybernetics, applying feedback principles to both engineering and biology. In the biological realm, early contributions by Claude Bernard on homeostasis and Walter Cannon's detailed work on physiological regulation in the 1930s established feedback as a cornerstone for understanding stability in living systems, bridging the concept into scientific and technological innovation.

Though there are desirable and undesirable negative feedback systems, we only discuss desirable ones here. Negative feedback systems operate by detecting deviations from a desired state or setpoint and initiating corrective actions to restore balance and stability. For instance, in engineering, these systems regulate outputs to ensure stability and precision. In missile guidance, negative feedback loops constantly measure the missile's position relative to its target, correcting its course when deviations are detected. Similarly, thermostats use feedback loops to adjust heating or cooling mechanisms when a room's temperature deviates from the setpoint, maintaining a stable environment (Figure 14.3).

Key Principles of Negative Feedback:

1. Error detection: Negative feedback systems monitor deviations from a desired state or goal.
2. Corrective action: Based on the detected error, the system applies adjustments to bring the output closer to the setpoint.
3. Dynamic stability: Negative feedback is crucial for maintaining equilibrium and preventing runaway effects in mechanical, biological, or social systems.

However, poorly designed feedback mechanisms, such as misaligned incentives in the cobra effect (Chapter 6), can lead to system instability if corrections are too aggressive or poorly timed.

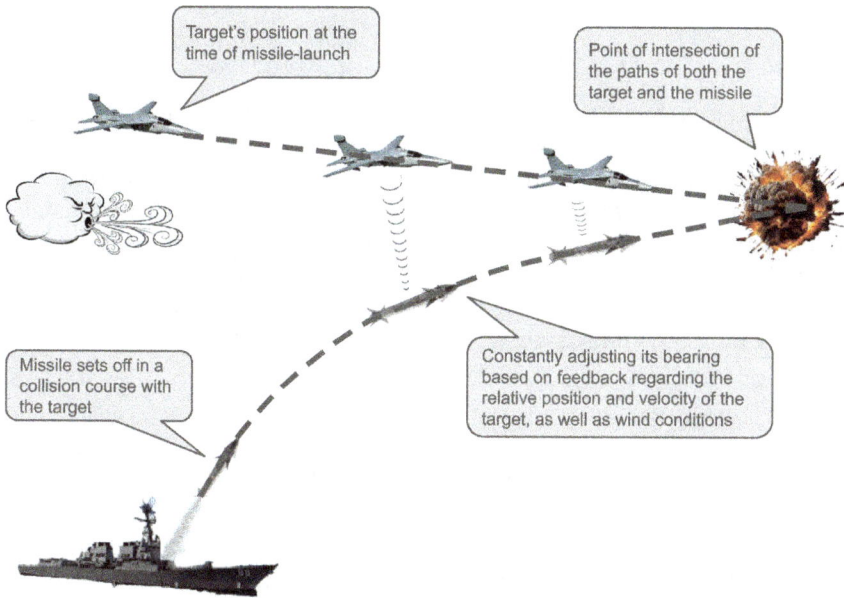

FIGURE 14.3
Overview of the missile feedback system.

In real life, complex systems operate through both positive and negative feedback mechanisms. In socioeconomics, for example, positive feedback loops drive phenomena like wealth accumulation, where the rich tend to get richer, and the poor tend to get poorer due to factors such as capital appreciation, preferential access to resources, and compounding advantages.

However, negative feedback mechanisms also exist to counterbalance these effects. Regardless of wealth, every individual is constrained by fundamental human limitations—we all have just one stomach, 24 hours in a day, a need for sleep, and a finite lifespan. As a result, the wealth gap is more meaningful when measured by lifetime expenditure rather than total assets, since the wealthy often act more as money managers than as direct consumers.

Additionally, wealth attracts adversaries, and the saying that "wealth rarely survives three generations" reflects a broader principle: the longer one remains in a privileged financial state, the more disruptive forces—social, economic, or even personal—accumulate against them. Moreover, happiness tends to follow a Weber–Fechner logarithmic law, rather than a linear function of wealth, meaning that beyond a certain point, additional riches bring diminishing returns in life satisfaction.

Together, these self-regulating forces form the negative feedback mechanisms that prevent unchecked wealth accumulation and contribute to the broader dynamics of socioeconomic balance.

14.6 Summary: Dualities as Intertwined Forces

This chapter explores the duality of evolution and devolution, challenging the conventional notion of "survival of the fittest." It examines the ethical and practical implications of balancing these forces to build sustainable and equitable systems. By analyzing positive and negative feedback mechanisms in system survival, the chapter integrates critical thinking and creative analogies to illustrate how evolution, devolution, and feedback systems function as complementary processes. Ultimately, it encourages readers to rethink progress, fairness, and sustainability in interconnected systems.

Key Takeaways

- **Duality across scales**: Evolution and devolution are interdependent forces that shape outcomes differently based on perspective and level of analysis.
- **Strategic implications**: Weaker players often survive longer by avoiding direct competition, showcasing the counterintuitive dynamics of survival in complex systems.
- **Balancing progress and equity**: A truly evolved society fosters fairness, ensuring that strength and dominance are not the sole determinants of success.
- **Positive feedback loops**: Self-reinforcing processes, such as network effects or speculative bubbles, drive exponential changes and growth.
- **Negative feedback mechanisms**: Stabilizing processes, like thermostats or homeostatic systems in biology, restore balance and prevent runaway effects.

Exercises

1. Define evolution and devolution in the context of interconnected systems. Provide an example of how these forces coexist within a biological or societal framework.
2. Explain why fitness is multifaceted and subjective. Use the example of the two couples with different health and fertility traits from the chapter to illustrate your point.
3. What critical thinking ideas and elements of creative analogies stand out to you in this chapter?

4. Analyze the shootout example (Andy, Bob, and Charlie) in terms of strategic decision-making. Why does the weakest player (Andy) have the highest survival probability in Scenario 2? Discuss how this dynamic applies to corporate or political competition.

5. Discuss how advances in medicine can lead to devolution at certain levels (e.g., weakening of the immune system). Propose strategies to mitigate such unintended consequences.

6. Extend the analogy of startups and tech giants to another field, such as sports or academia. How might weaker players benefit from the competition between stronger rivals?

7. Create your own analogy to explain the hierarchical interplay of evolution and devolution (e.g., cancer cells and organisms). Use a non-biological system, such as ecosystems or business organizations.

8. Debate whether equity in survival contradicts evolutionary principles. Should societies prioritize fairness over the traditional idea of "survival of the fittest"?

9. Discuss how the coexistence of evolution and devolution can impact long-term societal or technological progress. Are there instances where devolution might lead to positive outcomes?

10. Explain how statistical errors in drug development can lead to devolution, as described in the chapter. Suggest methods to ensure consistent progress in medical advancements.

11. Discuss the Cold War dynamics analogy. How does this reflect the coexistence of evolution and devolution in geopolitical strategies? Could similar dynamics apply to modern global challenges?

12. Identify a feedback mechanism in one field (e.g., biology, technology, and economics) and propose how its principles could be applied to solve a problem in another field.

15

What Constitutes Science and Scientific Evidence?

The world of wonder lies in understanding how evidence, causality, and creativity intertwine to shape science and innovation.

Note: The annotations for the AI-drawn illustrations in this chapter are provided to enhance understanding when reading Chapter 11.

15.1 The Multifaceted World

There is a widely held belief in an independent, objective world: a reality that exists regardless of human presence. Mathematics and physical science, for example, are thought to hold truths that are independent of human observers. Intelligent Martians and robots would presumably need to understand the same facts as humans. Similarly, a robot must accept that the world exists independently of itself and that it cannot fully comprehend every aspect of it.

However, Newton demonstrated that the eye acts as a lens, projecting an upside-down image of the world onto the retina. If all humans were color-blind except one, we might universally perceive the world as black and white, dismissing color as an illusion. Modern physics reveals additional limitations in our perception—we cannot see dark matter, ultraviolet light, or infrared radiation. Imagine beings with sensory perceptions vastly different from ours, such as a distinct sense of time or an ability to detect additional spatial dimensions. Would their perception define a different universe, perhaps a five-dimensional one? If superbeings exist, sensing far beyond what we can, does the universe expand in complexity as their sensory capacities increase? These questions suggest that the "objective world" is not singular or universal; it is instead observer-dependent, shaped by perception and intersubjective agreement. This extends our discussion on this topic in Chapter 13.

Our perceived world would differ dramatically if humans lived for only one second, one hour, one day, or a billion years. This multifaceted nature

DOI: 10.1201/9781003630081-15

of the world aligns with descriptions in quantum physics. Acknowledging the multifaceted objective world challenges conventional philosophy, where objectivity implies truth free from individual biases such as perception, emotion, or imagination.

In philosophy, objective truth refers to something true for everyone, independent of agreement or belief, while subjective truth is tied to individual experience. Over time, various theories of truth have emerged.

1. Correspondence theories assume the existence of an actual state of affairs and posit that true beliefs or statements align with that reality.
2. Constructivism contends that truth is not an external reality but is socially constructed, shaped by perception, convention, and historical context.
3. Consensus theories define truth as what is agreed upon by a specified group, whether all humans or a subset.
4. Pragmatic theories maintain that truth is validated through practical application and results.

From a logical perspective, a truth is what cannot, in principle, be proven false, not just what can be proven correct. However, truth and meaning rely on the language or tools used to describe them. Language definitions, in turn, rely on other words, creating a chain of meaning that ends when clarity is deemed sufficient or further pursuit is impractical.

The concept of intersubjective agreement highlights the role of shared understanding among conscious minds in defining truth. As Hilbe (1977) noted, truth emerges from collectively agreed-upon rules and conventions that describe facts within a shared conceptual framework. The absence of intersubjective agreement today underscores the difficulty in establishing common social truths.

The simulated-world hypothesis adds another layer of complexity, suggesting that reality could be a highly advanced simulation, indistinguishable from "true" reality. Conscious minds within such simulations might remain unaware of their simulated nature. Similarly, we cannot be sure that our remembered dreams align with their original experience. This uncertainty applies to waking reality: are we truly human, or are we creations of an advanced superintelligence? If humans eventually create human-like artificial intelligence (HAI) agents, it is conceivable that such beings might have already existed and, in turn, created us.

This multifaceted view of the world underscores the importance of using individual experiences—filtered through sensory perception, attention, and memory—to construct knowledge. Instead of relying solely on shared, commonsense knowledge, it advocates for the individualized patternization of experiences and responses as the basis for understanding.

15.2 Determinism, Freewillism, and Causality

> I do not believe in freedom of will. Schopenhauer's words, 'Man can indeed do what he wants, but he cannot want what he wants,' accompany me in all life's situations and reconcile me with the actions of others, even when they are truly painful to me. This awareness of the lack of free will prevents me from taking myself and others too seriously and helps maintain my sense of humor.
>
> *—Albert Einstein in My Credo (August 1932)*

Philosopher Benson Mates offers a similar perspective:

> Every event is the effect of antecedent events, which are themselves caused by events preceding them, and so on. Human actions are no exception to this rule. Although the causes of some actions are less understood than others, it's hard to deny that they have causes that determine their effects with the same certainty and inevitability found elsewhere. Specifically, those human actions usually called free are ultimately and inevitably the effects of events occurring long before the agent was born, over which they had no control. Since they couldn't prevent the causes, they couldn't avoid the effects. Consequently, despite appearances, human actions are no more free than the motions of the tides or the rusting of iron exposed to water and air.
>
> *—Benson Mates (1981, p. 59)*

Believers in free will acknowledge causal relationships but often stop at attributing certain events, like decision-making, to an agent's free will. They may not question further why free will makes a particular choice or why different individuals have different wills. However, if we continue asking *why*—the reason behind the reason, and so on—we will eventually reach a conclusion that free will, or a combination of all free wills (including animals'), is the origin of all outcomes. This raises deeper questions for free will advocates:

- Does free will exist since the beginning of the universe?
- Where does free will come from?
- If my free will is inherited from my parents (whose wills came from their ancestors) or imposed by an external entity, who is ultimately responsible for my actions, whether good or bad?

Saying "It happens because of free will" is unsatisfying from a scientific perspective. Researchers continuously ask why we make specific choices or what causes those choices. As neuroscience advances, understanding the human brain may leave little room for free will. Conversely, strict determinism—where every thought, belief, and action is preordained—can be equally unsettling. Under determinism, both committing a crime and punishing a criminal are machine-like actions, leaving no ethical dilemma.

For scientific purposes, causality between events A and B requires two key conditions:

1. Factor isolation laws hold:
   - If A occurs, then B occurs.
   - If A does not occur, then B does not occur.

2. Verifiability: The relationship must be persistent and, in principle, verifiable. This requires events in condition 1 to be repeatable.

   For a finding to become a scientific law, it must be independently verified. Repetition underpins the utility of causality, enabling us to predict and mitigate undesirable outcomes by addressing their causes.

   Our understanding of the universe relies on simplifications, as the human brain cannot fully encompass the vast complexity of the cosmos. Grouping similar events creates patterns and repetitions, enabling us to "discover" natural laws or casualties. In this sense, natural laws are part invention (through grouping) and part discovery (through pattern recognition). However, such approximations inevitably lead to exceptions for every law or causal relationship.

   This similarity-based grouping or categorization introduces random errors in statistical predictive models. For a believer in free will, the random error term also encompasses the component attributed to the unpredictability of free will.

   The mental grouping process, guided by the similarity principle, reinforces the belief that "everything happens for a reason." Yet, the inherent approximations ensure that no law or causal relationship is truly universal. We will explore this further in Chapter 16.

15.3 Science and Scientific Evidence

Philosophically, science can be understood as a systematic endeavor aimed at seeking "truth" by understanding, explaining, and predicting phenomena in the natural world through observation, experimentation, and reasoning.

From a logical perspective, "truth" encompasses more than what can be directly proved to be true. It also includes statements or propositions that cannot be proved false, especially within frameworks where proving truth or falsity may be limited by the system's axioms, available evidence, or computational limits.

Given the definition of truth, let us delve deeper into what constitutes supportive evidence for a hypothesis. To explore this, we will examine the topic through the lens of the raven paradox.

The raven paradox, or Hempel's paradox, is a classic problem in the philosophy of science that explores the nature of evidence and confirmation. The paradox can be stated as follows: Observing a black raven supports the hypothesis that "all ravens are black." By logical equivalence, observing a green apple should also support the hypothesis, since "all ravens are black" is logically equivalent to "all non-black things are non-ravens."

The raven paradox is based on contraposition in formal logic:

$$\forall x, \big(Raven(x) \rightarrow Black(x) \big) \Leftrightarrow \forall x, \big(\neg Black(x) \rightarrow \neg Raven(x) \big).$$

Thus, it is a paradox due to the fact that, while logically valid, the idea that a green apple supports the hypothesis about black ravens seems counterintuitive.

The paradox raises questions about what constitutes evidence for a general statement (scientific or legal).

In science, we often seek evidence to confirm or falsify hypotheses. The raven paradox illustrates that evidence confirmation is not always straightforward. For example, the hypothesis "all ravens are black" suggests that observing black ravens confirms it. Yet, by the paradox, observing non-black non-ravens (such as a red apple) also seems to confirm the hypothesis, as it does not contradict it.

This paradox underscores the need for careful selection of relevant evidence in scientific inquiry. It demonstrates that not all observations are equally useful for confirming or refuting hypotheses, which influences how scientists design experiments to focus on relevant data. Thus it can be used to guide the experiment design and data collections.

We now inspect the paradox through examples:

Suppose there are 10 items in a room. After you see five black (no other colors) ravens, you hypothesize; "all ravens are black in the room." The following question can be raised: if a green apple is observed later, does it support the hypothesis?

1. If the green apple is the last item observed in the room, then this observation provides definitive evidence or confirmation of the hypothesis. If the green apple is the ninth item observed in the room, then this observation provides very strong evidence supporting the hypothesis because the observation increases the probability of the hypothesis being true (e.g, from 8/10 to 9/10 dependent on your assumptions).

2. If a black raven is the ninth item observed in the room, then this observation provides the same level of evidence supporting the hypothesis as if it is a green apple. Furthermore, keep everything the same, but change the number of items to 100 or even to infinity in the universe. The conclusion should be the same in principle.

3. However, intuitively, why do we believe the statement "all ravens are black" more when we observed 1,000 ravens (all black) than when we just observed one black raven and 999 other items (green apple or other things) and thus believe observing a black raven is stronger direct evidence than observing a green apple? Do the colors of future ravens face the same uncertainty independent of what you have observed in the two scenarios?

4. When there are infinite objects or events, "observing a green apple" has diminished evidence to support the claim "all ravens are black."

We know the world is in a constant state of flux, and all predictive models will eventually become outdated. Yet, we continue to develop and rely on them. Why? By believing that history can statistically predict the future, we implicitly assume that changes occur gradually enough for us to detect, understand, and adapt to emerging patterns. This belief forms the foundation of our confidence in predictive models and underscores our reliance on data-driven decision-making. However, this does not eliminate controversies. For instance, one might argue, "the longer a stock rises, the more likely it is to rise tomorrow," while another might contend, "the longer a stock rises, the more likely it is to fall tomorrow because all trends eventually reverse." These conflicting interpretations highlight the inherent complexity and uncertainty in predictive reasoning.

15.4 The Art and Science of Causal Inference

Causal inference is the process of determining whether and how changes in one variable (the cause) lead to changes in another variable (the effect). It moves beyond correlation by establishing causality, often requiring a clear understanding of mechanisms, pathways, and assumptions. The ultimate goal is to predict outcomes under interventions, identify actionable causes, and distinguish spurious relationships.

Causal inference relies on assumptions such as exchangeability (no unmeasured confounders), positivity (sufficient variation in treatment or exposure), and no interference (one individual's treatment does not affect another's outcome). Directed acyclic graphs (DAGs) are often used to formalize assumptions about causal relationships.

Causal inference has broad applications across disciplines. In medicine, it helps evaluate treatment effects, such as the efficacy of new drugs. In policy, it is used to assess the impact of interventions like minimum wage laws. Public health relies on causal inference to identify risk factors for diseases, while machine learning increasingly incorporates causal reasoning to enhance predictive models.

15.4.1 Key Elements of Causal Inference

A cause is a variable that, when manipulated, results in a change in an outcome. Causes can be direct (e.g., smoking directly leading to lung cancer), indirect or root causes (e.g., matches leading to lung cancer through smoking), or operational, which are practical causes that can be controlled or acted upon for interventions.

A mediator is a variable that lies on the causal pathway between the cause and the outcome, explaining the mechanism of the effect. For instance, in the relationship between smoking and lung cancer, carcinogen exposure serves as a mediator.

A root cause initiates a causal chain that ultimately leads to an outcome but is not directly responsible for it. For example, matches lead to lung cancer indirectly by enabling smoking. Removing a root cause can disrupt the chain and reduce the outcome, but its impact depends on the context, such as whether alternatives like lighters are available.

A confounder influences both the cause and the outcome, creating a spurious association. Socioeconomic status, for instance, can influence both smoking and lung cancer. Properly adjusting for confounders is essential for isolating true causal effects.

A collider is a variable influenced by two or more causes. Conditioning on a collider can introduce spurious associations between its causes. For example, carcinogen exposure, which is influenced by both smoking and environmental pollution, can act as a collider.

15.4.2 Association and Causality

There are popular but over complex differences between association and causality in the literature. We are going to clarify them.

We often see the following five necessary criteria for causality:

1. Temporal precedence: the cause must occur before the effect.

2. Association: There must be a strong and consistent relationship between the two variables.

3. Elimination of confounding factors: It must be shown that no third variable (a confounder) is responsible for the observed relationship.

3. Plausible mechanism: A biological, physical, or logical mechanism must explain how the cause leads to the effect.

4. Replicability: The relationship should be observed consistently across different populations, contexts, and studies.

However, in my opinion, for criterion (3), there is always a possible confounding factor: For example, smoking can be considered as a confounder, while carcinogen is the direct cause of cancer. Criterion (4) depends on whether a plausible mechanism is subjective. Lastly, criterion (5) is met because criterion (2) implies a true association.

In fact, the difference between causality and association is in the additional requirement of temporal precedence in causality. Let us start with the deterministic criteria for association and causality. We use ⇒ for "imply" and ¬ for "negation."

Association is based on the principle of factor isolation:

1. If A ⇒ B and ¬A ⇒ ¬B, then A is associated with B.
2. If A ⇒ B and ¬A ⇒ B, then A is not associated with B.

Verified causality based on the principle of factor isolation:

1. Temporal Relationship: For A to cause B, A must precede B in time.
2. Principle of factor isolation:
   1. If A ⇒ B and ¬A ⇒ ¬B, then A is the cause of B.
   2. If A ⇒ B and ¬A ⇒ B, then A is not the cause of B.

If we agree on the deterministic sense of causality, then we can easily modify the definition to that in a statistical sense by adding "likely" in the statements.

15.4.3 Definition of Statistical Causality

1. Temporal relationship: A must precede B in time for A to be considered a potential cause of B.
2. Verified association based on the principle of factor isolation:
   1. If A likely ⇒ B and ¬A likely ⇒ ¬B, then A is likely associated with B.
   2. If A likely ⇒ B and ¬A also likely ⇒ B, then A is unlikely to be associated with B.

Here, the likelihood or probability can be calculated in different ways. This definition of causality offers a robust framework that bridges the gap between absolute and probabilistic reasoning. It aligns with the statistical nature of most real-world causal relationships while retaining the clarity of deterministic principles.

To elucidate key concepts surrounding causality and association, we present an example involving matches, cigarettes, carcinogens, and lung cancer (Chang, 2014):

Matches enable smoking, smoking causes carcinogen exposure, and carcinogen exposure leads to lung cancer. In this chain of events, matches can be replaced by lighters, while combusting tobacco can be replaced by e-cigarettes, which significantly reduce exposure to known carcinogens. Thus, the interpretation is context-dependent. This example highlights the importance of isolating key mechanisms when evaluating causality. By doing so, we recognize:

1. Smoking is a key causal agent for cancer due to carcinogen exposure, not because of nicotine or the act of inhalation.
2. Matches, lighters, or other ignition sources are replaceable and context-dependent enablers, not true causes.
3. When the mechanism changes (e.g., through e-cigarettes), the causal relationship to cancer also changes.

These examples clarify how causal factors can be necessary, sufficient, or replaceable within a causal chain. Matches and lighters are interchangeable enablers for smoking, while the mechanism (e.g., carcinogen exposure) is central to causality. Similarly, shifting from traditional cigarettes to e-cigarettes alters the causal dynamics, demonstrating that causality depends critically on mechanisms, but temporal or associative relationships remain unchanged (Figure 15.1).

In the chain of matches → smoking → carcinogen exposure → lung cancer, each step can have its own direct (proximate) and indirect (remote) causes, depending on the level of focus. Along the path of scientific discovery, the interpretation of these relationships may evolve. For instance:

- Initially, matches might be viewed as a direct cause of lung cancer.
- With the invention of lighters, smoking becomes recognized as the direct cause, and matches are reclassified as an intermediate cause.
- Later, with the discovery of carcinogens and the advent of e-cigarettes, carcinogens become the direct cause, smoking becomes an intermediate cause, and matches are relegated to the role of a remote cause.

In other words, any direct cause can eventually become a remote cause as scientific knowledge advances.

Lastly, given an effect like lung cancer, a factor such as matches can be interpreted either as a remote cause or a confounder, depending on the context.

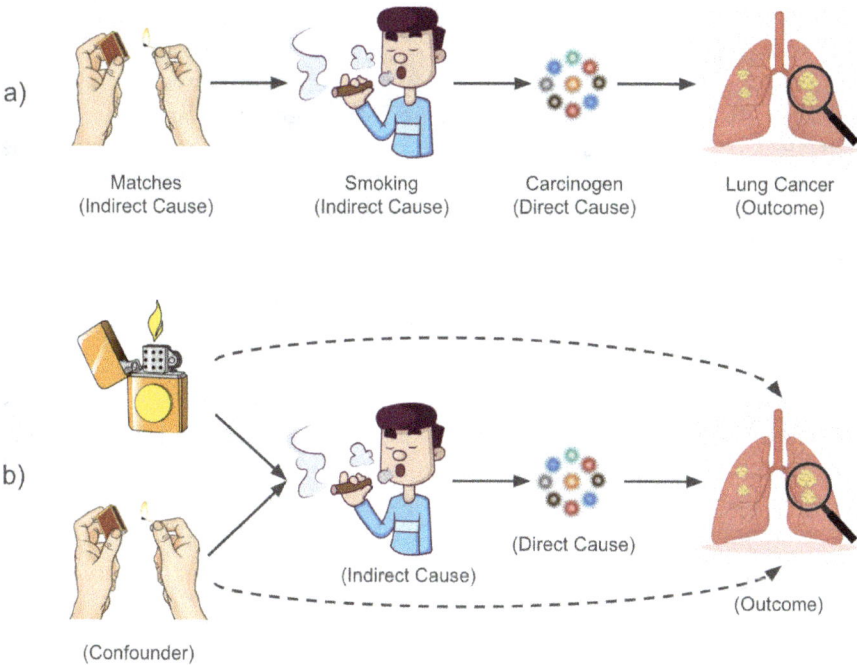

FIGURE 15.1
Exploring multiple interpretations of causal relationships.

1. Matches as a remote cause:
   - If matches are directly part of the causal chain (e.g., matches → smoking → carcinogens → lung cancer), they are a remote cause.
   - This happens when matches are necessary for lighting the cigarette in the observed context.
2. Matches as a confounder:
   - If matches are not part of the causal pathway because another factor (like a lighter) substitutes their role, matches become a confounder.
   - In this case, matches are still associated with smoking and lung cancer, but they do not lie on the causal path in this context.

The roles of variables—cause, mediator, confounder, or collider—are often subjective and context-dependent. For example, matches can be a root cause, a confounder, or a mediator depending on the assumptions and context, such as the availability of substitutes like lighters or e-cigarettes. Similarly, smoking can be viewed as a cause in one analysis and a mediator in another. These shifting roles highlight the flexibility and subjectivity inherent in causal inference.

Causality is an association, and similarly, an association is a causality as long as temporal precedence is met. In the example above, matches are a cause of lung cancer only as long as they are actively involved in the causal pathway through smoking. When their role in enabling smoking ceases, their association and causal relationship with lung cancer also end. This context-dependent perspective aligns with the mathematical definitions above and offers a clear, practical framework for understanding and analyzing causality.

15.4.4 Controversies in Causal Inference

Causal inference faces several important controversies, one of the most debated being the assumption of "no unmeasured confounders." This assumption simplifies analysis by presuming that all relevant confounders have been measured and accounted for. In reality, this assumption is almost always violated because unmeasured factors—potentially infinite, with or without names—will always exist as the sole source of response variability to treatments among different individuals or groups.

One argument is that the "no unmeasured confounders" assumption applies to the target population. However, if the target population is unique or homogeneous, the concept of unmeasured confounders becomes irrelevant because there is no variability to explain. Another argument is that the target population is heterogeneous, consisting of multiple subpopulations with different characteristics. If the heterogeneity is already addressed by measured factors in the model, unmeasured confounders may not be a concern. However, if the heterogeneity is defined by unmeasured factors, then unmeasured confounders are inevitable, making the assumption unrealistic.

Therefore, while the assumption of "no unmeasured confounders" serves as a practical starting point for causal inference, it has inherent logical flaws and is rarely realistic in practice. Researchers should approach causal inference with transparency, recognizing this assumption as conditional and supplementing analyses with sensitivity checks to account for potential biases from unmeasured confounders.

Second, the distinction between cause and association often depends on assumptions about causal structures, which allows the same factor such as matches to be interested as a cause or confounder under the same observations as we discussed above.

The focus on controllable causes also sparks debate. While controllable causes, such as smoking, are operationally meaningful for interventions, uncontrollable causes, like genetic predisposition, provide valuable insights into mechanisms and risk stratification. The controllable causes can evolve as science and technology develop.

Recursive causality introduces additional challenges. Estimating probabilities for potential outcomes often require intermediate inferences, which can compound uncertainty and depend heavily on model assumptions. For

instance, estimating smoking's effect on lung cancer may first require estimating carcinogen exposure due to smoking.

The subjectivity inherent in constructing causal diagrams introduces flexibility but can lead to biased or inconsistent interpretations. Transparency
about assumptions is critical to mitigating this concern.

15.4.5 Steps in Causal Inference

Causal inference begins with defining the causal question, clearly articu-
lating the treatment, outcome, and potential confounders or mediators. For
instance, one might ask whether smoking cessation reduces lung cancer risk.
Next, researchers construct a causal diagram, such as a DAG, to visually represent assumed relationships between variables and identify confounders,
mediators, colliders, and pathways.

Estimation methods depend on the study design. Experimental approaches,
like randomized controlled trials, are preferred when feasible. Observational
methods, such as matching, propensity scores, instrumental variables, or
regression, are employed when experiments are not possible.

Assumptions are checked to ensure exchangeability (treated and untreated
groups are comparable), positivity (all treatment levels have non-zero probabilities), and no interference. Results are then interpreted in context, with sensitivity analyses used to evaluate robustness under alternative assumptions.

15.4.6 Conclusion

Causal inference provides a structured framework for understanding and
intervening in cause–effect relationships. Despite its subjectivity and reliance
on assumptions, it disentangles complex interactions, identifies actionable
causes, and predicts outcomes under interventions, making it indispensable
across scientific disciplines.

Causality is as much an invention as it is a discovery. While causal inference provides tools to construct and reason about cause–effect relationships,
these relationships are shaped by human perspectives, goals, and assumptions. At the same time, alignment with empirical evidence and mechanisms
bridges the gap between invention and discovery, making causality both a
tool for understanding and a guide for action.

15.5 The Paradox of Discovery

Science is deeply rooted in discovery and creativity, while technology transforms these discoveries and creative ideas into inventions and innovations.
This relationship highlights the interplay between understanding and

application. A question often arises: Can AI have creativity? To address this question satisfactorily, we must explore the nuanced meanings of "discovery," "creativity," and "innovation."

Generative AI, by utilizing posterior probability distributions, and evolutionary AI, through techniques such as genetic algorithms and genetic programming (as discussed in Chapters 1 and 18), can generate new data, creative ideas, or even inventions. A notable example is AlphaFold, which has demonstrated the ability to discover and invent new molecular structures, advancing the boundaries of biology and chemistry.

In the realm of entertainment and art, AI's creativity becomes evident. Consider the following poem, created collaboratively by ChatGPT-4o and myself, which reflects on the dynamic interplay between critical thinking and creative analogy (Figure 15.2).

The poem is translated into Chinese. Translating from one language to another serves as an analogy, mapping ideas, structures, and meanings between distinct linguistic systems. Both processes aim to preserve core relationships while adapting to the nuances of the target system. This approach not only highlights similarities but also addresses unique differences, mirroring how analogies draw parallels between distinct concepts.

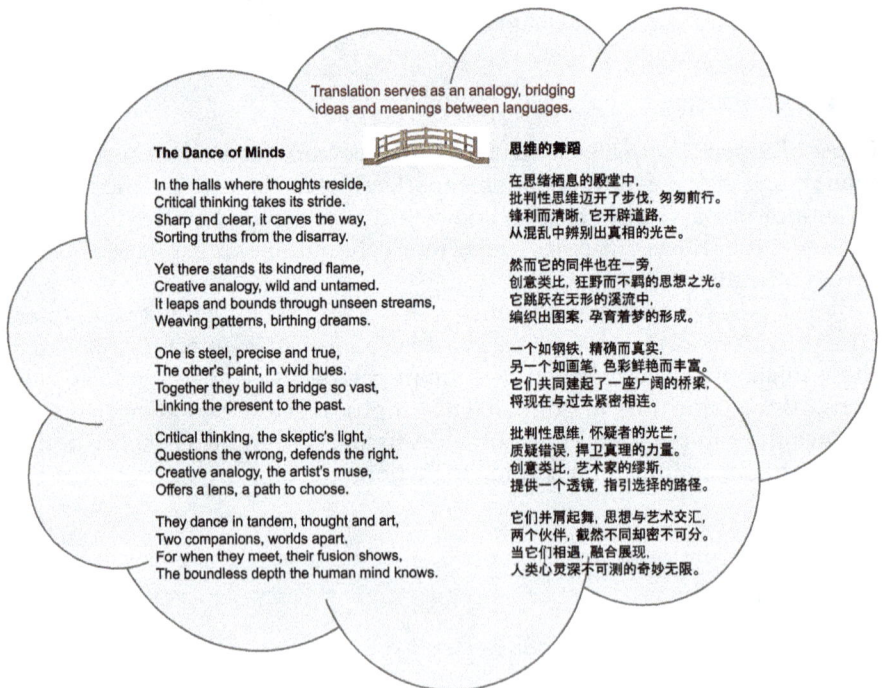

FIGURE 15.2

Critical thinking and creative analogies: a dynamic interplay.

Discovery, paradoxically, often blurs the line between the discoverer and the interpreter of the discovery. This interplay raises profound philosophical and cultural questions about attribution, intention, and the evolving meaning of what we call "discovery."

15.5.1 Historical Interpretation beyond Intent

As history unfolds, interpretations of discoveries, texts, and artworks often transcend their creators' original intentions. For instance, readers analyzing historical texts or novels frequently project contemporary concerns onto works created in the past, offering explanations that the authors may never have envisioned. Similarly, the meaning of an artwork can evolve, with modern viewers experiencing it in ways that diverge significantly from or even surpass the artist's initial intent. This phenomenon reveals a paradox: the interpreter, not the creator, often shapes the enduring legacy of a discovery or creative work.

15.5.2 Cultural Overinterpretation

A striking example comes from the Cultural Revolution in China (1966–1976), where Mao Zedong's poetry was subject to overinterpretation, elevating it to levels of significance far beyond its literary merit. These interpretations were shaped by political necessity rather than artistic intention, amplifying the perceived quality of the work tenfold. This phenomenon demonstrates how external contexts can transform the reception of a discovery or creation, making it something it may never have been meant to be.

15.5.3 Scientific Discovery and Application

In science, the interpretation and application of discoveries often lead to further breakthroughs. Analogies drawn from an initial discovery can open doors to unforeseen possibilities. For example, a conjecture, whether proven correct or incorrect, can serve as a catalyst for scientific advancement. Fermat's last theorem, though famously unproven during Fermat's lifetime, generated centuries of mathematical exploration and eventually led to profound developments in number theory. The theorem itself became less important than the golden egg it laid—the transformative research it inspired.

15.5.4 AI and the Attribution of Discovery

These questions become even more complex in the context of AI. If an AI agent were to generate the exact text of Darwin's *On the Origin of Species* before Darwin, would it qualify as the discoverer of evolution? Or does the act of recognition and interpretation by a human confer the status of discovery (Chang, 2014, 2023)?

In some respects, this mirrors the role of randomness. A random text generator, given infinite time, could theoretically produce Darwin's text. If a human reader identifies the text's significance, is the discovery attributable to the random generator or to the human who discerned its meaning? Similarly, should the explanation of AI-generated text differ from Darwin's original explanation simply because the origin is mechanical rather than biological? These questions challenge our understanding of creativity and attribution.

The paradox of discovery highlights the interplay between creation and interpretation, reminding us that discoveries do not exist in isolation. They are dynamic, evolving entities shaped by the contexts in which they are understood. Whether through cultural reinterpretation, scientific application, or even AI, discovery transcends the boundaries of its origin, becoming a collective human endeavor.

In Chapter 19, we will delve deeper into how to equip humanized AI with the capabilities of creativity, intentionality, and scientific discovery. This exploration will reveal pathways to transcend the randomness of generative processes and imbue AI with a deeper sense of purpose and innovation.

15.6 Frequentist and Bayesian Concepts of Probability

The overarching goal of science is to seek the truth—not in an absolute sense, but within a probabilistic framework. This pursuit is grounded in two distinct probability paradigms: Bayesian probability (BP) and frequentist probability.

BP is often described as a "degree of belief," but this term can be overly arbitrary and potentially more confusing than the concept of probability itself. For instance, "degree of belief" could theoretically be defined using various functions, such as f (the proportion of desired outcomes), f, f^2 or $\dfrac{2f}{1+f}$, among others. All these functions share the property of ranging from 0 (representing events that never occur) to 1 (representing events that always occur). However, the choice of function is subjective, which underscores the ambiguity and lack of standardization in defining "degree of belief," much like defining "hotness" by temperature, the logarithm of temperature, or other forms.

Frequentist probability defines probability as the long-run relative frequency of an event occurring in repeated, independent trials under identical conditions. For example, the probability of flipping heads with a coin is understood as the proportion of heads observed over a large number of flips.

I wonder: are the frequentist and Bayesian concepts of probability as fundamentally different as apples and bananas? If so, why don't we use distinct terms to avoid confusion in daily conversations and scientific research?

FIGURE 15.3
Bayesian vs. frequentist perspectives: different meanings of the same term.

Perhaps the distinction becomes clearer when we view BP as an evolving estimate of Frequentist probability over time as data accumulates. Bayesian inference updates beliefs dynamically with new data and often converges toward the long-run frequencies described in the Frequentist framework. This perspective reconciles the two views, highlighting their complementary roles in understanding and interpreting probability (Figure 15.3).

Here is how this works in practice:

1. Convergence with data:
   - As more data is collected, the Bayesian posterior distribution for the probability parameter p (e.g., probability of heads in a coin flip) typically becomes more sharply peaked around a single value. This peak is often close to the frequentist estimate, such as the sample proportion in the case of a coin flip.

- This convergence reflects the idea that, with enough data, the Bayesian estimate of p will increasingly align with what a frequentist would expect as the long-run frequency of heads.

2. Prior influence diminishes over time:

   - In Bayesian inference, the choice of prior can significantly affect the posterior distribution, especially with limited data. However, as more data accumulates, the influence of the prior diminishes, and the posterior distribution is increasingly determined by the observed data alone.

   - When the prior becomes effectively negligible, the Bayesian posterior essentially reflects what would be a frequentist's probability estimate based on the accumulated data.

3. BP as a dynamic estimate:

   - You can think of BP as a dynamic, evolving estimate of the true underlying probability. At each stage, BP reflects current beliefs about p, combining prior information with observed data.

   - With more data, this evolving estimate refines itself, often approaching the frequentist notion of probability as an objective long-run frequency.

4. Frequentist probability as a limiting case:

   - In the limit of infinite data, Bayesian posterior distributions become very narrow, essentially "collapsing" around the frequentist probability estimate. This limiting behavior is why Bayesian methods can be seen as providing a refined, iterative estimate of the frequentist probability as new evidence accumulates.

5. Prior knowledge usage

   - Frequentists incorporate prior knowledge in experimental design and in identifying the relevant target population. However, they do not use prior knowledge when determining probabilities of, for example, the treatment effect.

   - Bayesians, on the other hand, use prior knowledge in both experimental design and data analysis, incorporating it to determine the posterior probability of the treatment effect.

In some simple cases, BP can indeed be viewed as an estimate of frequentist probability that refines itself with each new data point. Over time, as data accumulate, the Bayesian perspective can converge toward what a frequentist might consider the "true" probability, making it feel like a dynamic pathway to the frequentist probability interpretation. This convergence, however, depends on the assumption that the Bayesian approach uses a reasonable prior and that the model aligns well with the data-generating process.

To reduce confusion, I hope the statistics community universally adopts the frequentist definition of probability while recognizing both Bayesian and frequentist approaches for estimating probabilities.

While this specific paradox may not always be highlighted explicitly, the tension between the Bayesian and frequentist schools and the interdependence of their methods has been a long-standing topic of debate in the statistical community.

Edwin Thompson Jaynes, a prominent Bayesian statistician, recognized that frequentist methods can be useful for determining objective priors or for deriving likelihood functions. However, he argued that these methods should still be considered within the Bayesian framework, where probability is interpreted as a degree of belief rather than a frequency. While Jaynes championed the Bayesian approach, he acknowledged that frequentist tools could provide empirical data to inform Bayesian priors, which reflects the subtle reliance of Bayesian methods on frequentist ideas.

Andrew Gelman, a prominent Bayesian statistician, has pointed out that Bayesians often use frequentist ideas for model checking, calibration, and in setting prior distributions. In their book, *"Teaching Statistics: A Bag of Tricks,"* Gelman and Nolan acknowledge that in practice, statisticians frequently borrow ideas from both the Bayesian and frequentist traditions, and that the strict dichotomy between the two is often more philosophical than practical. This reflects the recognition of the interplay between the two approaches, where frequentist methods are used even within a Bayesian paradigm, demonstrating the pragmatic integration of methodological traditions.

Bradley Efron, a leading statistician, has written about the strengths and limitations of both Bayesian and frequentist methods. In his well-known paper, *"Why Isn't Everyone a Bayesian?"*, Efron discusses the advantages of frequentist methods for certain tasks, such as hypothesis testing and objective data analysis, while also highlighting situations where Bayesian methods offer clear advantages, particularly in incorporating prior information. He acknowledges that in practice, many statisticians use a hybrid approach, employing frequentist tools within a Bayesian framework. This pragmatic approach, which blends elements from both schools of thought, underscores the ironic relationship where the methods are interdependent despite philosophical disagreements.

The empirical Bayes method, developed by Herbert Robbins and further popularized by Bradley Efron, is a classic example where frequentist principles are used to estimate the prior distribution in a Bayesian analysis. This approach explicitly blends frequentist and Bayesian ideas, reflecting the interdependence of the two schools. Empirical Bayes methods illustrate the irony, as they use frequentist data to inform Bayesian priors, thus acknowledging the utility of frequentist concepts within a Bayesian framework.

In simple cases, such as binary outcomes (where probability itself is a model parameter), frequentist and Bayesian probabilities share a common destination, called "rooted probability": the proportion of favorable events with an infinite sample size or data. However, their paths to this goal differ as data accumulates over time. Frequentists rely solely on well-defined data, while Bayesians incorporate both well-defined data and additional information, which may be ill-defined (objective Bayesian priors) or even undefined (subjective priors). Interestingly, one might argue that both objective and subjective Bayesian priors are ultimately rooted in direct or indirect experiences passed down through genes from parents.

This shared destination provides a foundation for meaningful debate about the concept of probability, while the differing paths make the debate itself contentious. Evaluating which approach is better in specific situations depends on how closely each aligns with the common destination.

Although frequentist and Bayesian methods can eventually converge to the same value as data grows, a fundamental challenge persists: establishing a shared connotation for "rooted probability." Under Kolmogorov's axioms, the definition of rooted probability is not unique, leaving room for interpretation and continued debate.

In summary, the interdependence between Bayesian and frequentist methods has been acknowledged by numerous statisticians and scholars. The irony lies in the fact that while Bayesians often critique frequentist interpretations of probability, they frequently incorporate frequentist-derived probabilities into their analyses in a quiet, pragmatic manner. This blending of approaches is evident in discussions about model checking, prior selection, and empirical Bayes methods, demonstrating that, in practice, the two paradigms are more complementary than contradictory.

15.7 Summary: The Nature of Science, Evidence, and Discovery

This chapter explores the foundations of science and scientific evidence, examining how causality, evidence, and creativity intertwine to advance knowledge. It critiques deterministic and free will paradigms, clarifies the distinction between causality and association, and emphasizes the role of creativity and innovation in scientific discovery. Blending critical thinking with creative analogies, it challenges conventional views on causality, evidence, and discovery, fostering a nuanced understanding of science's dynamic and creative nature. This chapter also highlights the multifaceted nature of reality, shaped by sensory limitations and intersubjective agreements, while addressing the interplay between Bayesian and frequentist probability concepts.

**Key Takeaways from Critical Thinking and
Creative Analogical Perspectives**

1. Philosophical foundations: Determinism and freewillism are contrasted, revealing how they influence our understanding of causality. The multifaceted nature of the world underscores the importance of recognizing observer-dependent realities shaped by perception and agreement.

2. Discovery beyond origin: Scientific discoveries are dynamic, shaped by creators, interpreters, and future applications, with AI and generative models increasingly contributing to creative processes.

3. Causality and evidence: This chapter emphasizes the need for temporal precedence and the principle of factor isolation to establish causality.

4. Analogies in causality: The relationship between matches, smoking, and lung cancer exemplifies how causal roles can shift with evolving knowledge, illustrating the importance of isolating mechanisms in understanding causality.

5. Probabilistic reasoning: Bayesian and frequentist approaches to probability are presented as complementary, highlighting their practical convergence while retaining philosophical differences.

Exercises

1. Explain how perception shapes our understanding of the "objective world." Provide an example of how different sensory perceptions (e.g., humans vs. other species) might result in distinct realities.

2. Is the multifaceted nature of the objective world a result of the limitations of our sensory organs, as no unique world can be defined through the evolution of sensory perception, or is it inherently unique? Will we, or some super beings, eventually be able to perceive this uniqueness?

3. Compare determinism and free will using examples from this chapter. How do these philosophies influence our understanding of causality?

4. What critical thinking ideas and elements of creative analogies stand out to you in this chapter?

5. Analyze the raven paradox. Why does observing a green apple seem to support the hypothesis "all ravens are black"? Discuss its implications for scientific evidence and experiment design.

6. Differentiate between causality and association using the example of smoking, carcinogens, and lung cancer. How can confounders and mediators affect our interpretation of causal relationships?

7. Develop an analogy to explain the difference between Bayesian and frequentist probabilities. Use a real-world context, such as weather forecasting or medical testing.

8. Create an analogy that demonstrates the interplay between discovery and interpretation in science. Use historical or modern scientific examples to illustrate your analogy.

9. Debate whether truth is better defined through objectivity (correspondence with reality) or subjectivity (individual perception). Support your argument with examples from this chapter.

10. Discuss the ethical implications of causal inference in fields like medicine or policy. How should researchers balance the need for actionable causes with the complexities of causality?

11. Explore the simulated world hypothesis presented in this chapter. What philosophical and scientific questions does this hypothesis raise about the nature of reality?

12. Discuss how Bayesian and frequentist approaches complement each other in scientific research. Provide an example of a situation where integrating both methods would be beneficial.

16

The Similarity Principle

The similarity principle serves as the cornerstone of all scientific disciplines.

The similarity principle unites diverse scientific fields by highlighting common patterns and structures.

The effectiveness of analogies in science stems directly from the similarity principle.

16.1 The Art of Similarity Grouping

16.1.1 Revelation from Simpson's Paradox

When making scientific decisions or even choices in daily life, we often begin by asking: Who or what constitutes the relevant data source? In other words, we define the population from which we will draw our observations. A sample space encompasses the minimum to maximum values that the measure could realistically take within the target population. For example, in clinical trials, this process is formalized by defining the target patient population through inclusion and exclusion criteria specified in the clinical trial protocol. These criteria determine who is eligible to participate and ensure the study includes the most relevant individuals to answer the research question.

Choosing the target population in a clinical trial is a critical step that ensures the study addresses its scientific objectives while being ethical, feasible, and generalizable. The target population refers to the group of individuals who will be eligible to participate in the trial based on predefined characteristics. The process involves carefully balancing scientific and practical considerations.

The determination of target population or relevancy is based on prior information; thus, somewhat subjective and controversies are unavoidable as illustrated by the well-known Simpson's paradox.

Suppose we have two options for conducting clinical trials to investigate the efficacy and safety of a test drug: (1) a single large trial with both male and female patients and (2) two separate trials for male and female patients conducted sequentially. The same drug is tested on the same patient populations, resulting in identical data. However, the data is analyzed differently:

DOI: 10.1201/9781003630081-16

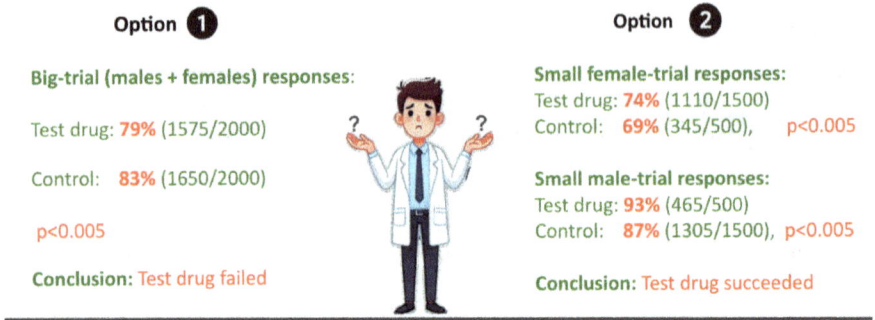

FIGURE 16.1
Simpson's paradox in clinical trial decision-making.

either aggregated (Option 1) or separated by trial (Option 2). Here are the hypothetical results (Figure 16.1):

Option 1: A single large trial with both male and female patients.
- Test drug response rate: 79% (1,575/2,000).
- Control response rate: 83% (1,650/2,000).
- Result: $p < 0.005$.
    Conclusion of option 1: The test drug is significantly worse than the control and fails.
Option 2: Two separate trials for female and male patients conducted sequentially.
- Female trial: Test drug response rate=74% (1,110/1,500), control response rate=69% (345/500), $p < 0.005$.
- Male trial: Test drug response rate=93% (465/500), control response rate=87% (1,305/1,500), $p < 0.005$.
    Conclusion of option 2: The test drug is superior to the control for both male and female patients.

This paradox demonstrates how conclusions can differ depending on whether data is analyzed in aggregate or by subgroup. Further subdivision of the target population based on attributes such as age could also yield different conclusions.

We will explore how to resolve or avoid this paradox using similarity-based machine learning (SBML).

16.1.2 Paradox of the Traffic Ticket

Let me share a personal story. When I appeared in court for my first traffic citation, I argued, "I don't believe this was my fault. I've been driving for

13 years without any traffic violations." Essentially, I was making a Bayesian argument: based on my strong prior record, in all past situations where I was involved in car accidents, none were my fault. Therefore, it was very likely that this incident was not my fault either.

The judge, however, responded kindly, "Yes, I understand, but everyone has a first time." Her perspective was based on a very different prior. She likely thought that the (unconditional) probability of a driver with a 13-year history eventually having at least one traffic violation is quite high. This difference in priors highlights how subjective they can be, as they depend on how individuals interpret and apply their knowledge.

In this case, one could use traffic records from drivers with similar characteristics (age, gender, driving history, etc.) to construct a prior. However, the definition of "similar" is itself subjective. This raises an important question: is it fair to judge my actions based on what others with similar characteristics have done (i.e., the prior)?

At the same time, is the essence of science not to make inferences based on prior knowledge and observed patterns in nature? The tension here lies in balancing fairness in individual cases against the generalizations derived from prior knowledge. What do you think?

16.1.3 Paradox of Life Expectancy

A Chinese man once traveled from China to Japan. Upon landing, he excitedly announced that he had just increased his life expectancy by 11.3 years—citing the fact that life expectancy in Japan is 86.1 years compared to 74.8 years in China. His reasoning was amusingly straightforward: while in China, he considered himself Chinese and calculated his life expectancy based on the average for all Chinese people. However, upon arriving in Japan, he adopted the perspective of a Japanese resident and recalculated his life expectancy accordingly.

This playful shift highlights how our perceived "causal space" or target population can change depending on where we are or how we identify at a given moment. However, speaking for myself, I do not recall feeling any younger when I stood in the Massachusetts State Hall during the Naturalization Oath Ceremony and officially became a U.S. citizen!

16.1.4 Different Similarity Grouping in Bayesian and Frequentist Paradigm

Similarity grouping is the basis for calculating the probability in Bayesian and frequentist paradigm. In practice, defining repeated experiments involves subjective similarity grouping because real-world events or trials are never perfectly identical. The decision to treat certain experiments as "repeated" is based on human judgment about which factors are important to control and which can be considered irrelevant or random noise. This subjectivity underpins much of how we design experiments and interpret probabilistic results in both frequentist and Bayesian frameworks.

Similarity grouping and the diminished effect in larger categories: The way conditions are grouped or divided significantly influences how their impact is perceived. Politicians, policymakers, or interest groups often manipulate such categorizations to either amplify or downplay the significance of diseases based on their agendas.

Understating impact by splitting categories: Dividing a broad category like cancer into smaller subcategories—such as breast cancer, lung cancer, and skin cancer—reduces the apparent significance of each. This fragmentation can make each category seem less urgent, reducing public attention and justifying smaller-scale funding or interventions. For example, focusing on rare or less common cancers may give the impression that the overall burden of cancer is minimal, potentially leading to delays in broader healthcare initiatives.

Overstating impact by grouping categories: Conversely, grouping multiple diseases under a single umbrella can magnify their collective importance. For instance, combining all types of cancer or grouping cancer with other chronic diseases like heart disease and diabetes inflates the overall case numbers or mortality rates. This approach is often used to justify significant funding or large-scale healthcare reforms. While it draws attention to the issue, it can overlook the unique causes, treatments, and priorities of individual diseases.

Ethical implications and political manipulation: This selective framing poses ethical concerns. Manipulating data to either fragment or consolidate disease categories can misallocate resources, favoring politically advantageous areas over genuine public health needs. For example, emphasizing one group of diseases while neglecting others may result in overfunding some areas while leaving critical gaps in others. This strategy shapes public opinion and drives policies that may not align with the actual needs of patients.

Thus, the way health conditions are grouped or divided is a powerful tool that can shape public perception and policy decisions. Whether by overstating or understating the significance of diseases, such manipulations can have profound consequences on resource allocation, healthcare priorities, and public health outcomes. Ethical use of data and thoughtful categorization are essential to ensure policies are based on genuine needs rather than other reasons.

16.2 The Similarity Principle and Applications

16.2.1 The Similarity Principle

Albert Einstein suggested that the grand aim of science is to explain the widest range of empirical facts through logical deductions from the fewest hypotheses or axioms. Science studies recurring events, seeking to uncover causal relationships in which one event serves as the cause and another as the effect. Through studying history, we aim to identify patterns and laws to predict future outcomes when similar causes recur. Although no two events are identical, science groups similar, one-time occurrences to create the appearance of recurring events (Chang, 2014).

Karl Popper argued that science requires hypotheses with testable predictions. Scientific theories are continuously tested against new observations, with new paradigms emerging when they solve problems more effectively. However, testing every possible cause–effect relationship is impractical; therefore, our brains group events by similarity, as they cannot process each event individually.

Thus, causality is not an absolute fact but an interpretation based on recurring conjunctions of similar events. Scientific inquiry involves simplifying historical complexity by identifying patterns and deducing laws from these recurring relationships. The concept of causality relies on the idea of recurrence and the definitions of "same" or "identical." Yet, no event recurs exactly; real-world events are approximations, where certain details are ignored, making this grouping process inherently subjective.

The foundation of prediction lies in causality, rooted in the similarity principle: similar situations will likely result in similar outcomes. When we claim two situations are "the same," we overlook subtle differences that may eventually contradict the scientific law we believe in. When these contradictions emerge, the law may need revision. Both science and statistics account for exceptions—science reacts when exceptions arise, while statistics anticipates variability.

Scientific inference aims to approximate causal truths that our brains can process. By grouping events based on likeness and ignoring differences, we can observe recurring patterns in causes and effects, forming the basis for understanding causal relationships. However, this implicit grouping is controversial, as ignored differences can impact how we define "same events." These variations lead to differing interpretations of similarity, influencing how we construct causal relationships.

Chang's idea, articulated in his works *Principles of Scientific Methods* (2014) and *Foundation, Architecture, and Prototyping of Humanized AI* (2023), highlights the similarity principle as the fundamental basis of analogical reasoning: Similar things will behave similarly, and the more similar they are, the more similarly they behave.

This principle encapsulates the essence of analogies by emphasizing the predictive and explanatory power that emerges from recognizing patterns of similarity.

16.2.2 Applications of the Similarity Principle

The similarity principle—the idea that similar conditions or entities are likely to yield similar outcomes—manifests in various aspects of life, science, and decision-making. Here are several examples:

1. **Education and Development**
   - Age-based grouping in schools: Children of the same age are grouped into the same grade because they are presumed to have similar developmental stages and learning abilities, making it easier to design age-appropriate curricula.
   - Peer influence on academic choices: If friends or relatives attend a prestigious university (e.g., Harvard) and achieve success, parents often want their children to follow a similar path, expecting comparable outcomes.

2. **Historical and Seasonal Patterns**
   - Terrorism risk on specific dates: Events like September 11th are associated with higher vigilance due to past attacks on the same day. The similarity in the date (month and day) leads to patterns of heightened security.
   - Seasonal events and patterns: Certain weather events, such as hurricanes, are expected during specific months due to historical data showing their seasonal recurrence.

3. **Behavior and Social Influence**
   - Predicting outcomes from behavior: If drunk driving frequently leads to accidents, the similarity principle suggests that any drunk driver is likely to cause an accident, reinforcing stricter laws and campaigns.
   - Friendship and social circles: People often mimic the career, education, or lifestyle choices of their peers, believing that similar paths will yield similar successes.

4. **Consumer and Marketing Decisions**
   - Brand loyalty: Consumers assume that if one product from a brand meets their expectations, other products from the same brand will perform similarly (e.g., buying multiple Apple devices).
   - Recommendations: Platforms like Amazon or Netflix suggest products or content based on user preferences, relying on patterns of similarity between users and their behaviors.

5. Medicine and Health

- Risk prediction: If a specific genetic mutation predicts a higher likelihood of disease, individuals with similar genetic profiles are assumed to face comparable risks.
- Treatment effects: Drugs that work well for patients with certain biomarkers are expected to yield similar benefits for other patients with the same biomarkers.

6. Law and Policy

- Recidivism predictions: Criminal justice systems often assume that individuals with similar criminal records are more likely to reoffend, influencing sentencing or parole decisions.
- Policy success: If a policy worked in one country or state, it is often implemented in others with similar socioeconomic conditions, assuming comparable success.

7. Personal and Emotional Decisions

- Parenting practices: Parents replicate strategies that worked for their first child with their subsequent children, expecting similar outcomes.
- Happiness and relationships: People gravitate toward positive, happy individuals, believing their happiness will similarly influence their own well-being.

8. Risk Assessment and Insurance

- Actuarial predictions: Insurance companies use similarity-based actuarial models to assess risk, such as setting higher premiums for individuals in high-risk demographics (e.g., young drivers or smokers).
- Safety policies: Workplace safety measures are implemented based on patterns of accidents in similar environments.

9. Scientific and Research Applications

- Model organisms: Mice are used in biological research because their genetic and physiological similarity to humans often predicts comparable responses to drugs or treatments.
- Astronomical predictions: Scientists assume that similar stars or galaxies exhibit similar behaviors, allowing for extrapolation in studying the universe.

10. Cultural and Behavioral Norms

- Traditions and rituals: Certain actions or practices are repeated because they are believed to yield favorable outcomes, such as wearing specific colors for good luck.
- Copycat phenomena: High-profile events, like protests or crimes, often inspire similar actions elsewhere due to perceived similarities in circumstances.

The similarity principle teaches us that our happiness is closely linked to the happiness of those closest to us. By surrounding ourselves with positive, joyful people and contributing to their well-being, we create lasting and meaningful relationships. This mutual exchange of happiness has a profound impact, far greater than focusing on distant or abstract relationships.

The world is vast, but we are small. Instead of trying to "save the world" or bring happiness to everyone, it is more practical—and impactful—to focus on the people around us. True kindness begins at home, with those we interact with daily. Pretending to know everything or attempting to change the entire world is not an act of great love; it often reflects arrogance.

In essence, our happiness stems not from trying to fix faraway problems but from nurturing the well-being of those within our immediate reach. Treating those around us with care and fostering positivity is the most effective and authentic way to create a ripple effect of happiness.

In conclusion, the similarity principle is central to scientific discovery, prediction, and decision-making. We inherently rely on it: Similar conditions often lead to similar outcomes. Without this principle, analogies would collapse, and scientific progress would stagnate. However, defining similarity is not without its challenges. Claiming that two things are similar simply because they produce similar outcomes introduces a form of circular reasoning—a restatement of the principle in reverse. While the similarity principle is indispensable for simplifying complex problems and anticipating outcomes, its application requires caution to avoid oversimplification or ignoring key differences, which could lead to flawed conclusions.

16.3 Similarity-Based Machine Learning

While the similarity principle is used instinctively in daily life, science, and statistics, SBML makes it operational by enabling its application in learning processes. For instance:

- Ligands with similar structures will behave similarly or have a similar mechanism of action.
- The effectiveness of a drug on patients can be somewhat predicted based on its effects on animals.
- Patients with similar disease profiles, gender, and age are expected to exhibit similar responses to the same medical interventions.
- Greater similarity across multiple attributes leads to more consistent responses.

16.3.1 Similarity Measures

A similarity measure quantifies the resemblance between two subjects (e.g., individuals, objects, or event sequences). It is typically an inverse function of dissimilarity or distance (d).

For example, the exponential similarity function S is defined as:

$$S = \exp(-d),$$

where d is the distance between two subjects, ranging from 0 (identical subjects) to infinity (completely different subjects), with the corresponding S ranging from 1 (maximum similarity) to 0 (no similarity). The distance between the j-th subject to the subject to be predicted is

$$d_j = \left| R\left(X - X_j\right) \right|,$$

where R is a row vector of attribute-scaling factors that weigh each attribute's importance and X and X_j are attribute vectors of the new and j-th subjects, respectively.

16.3.2 Similarity-Based Learning

According to the similarity principle, to predict the outcome Y for a new subject with attributes X, SBML uses similarity-weighted outcomes from N training subjects. Mathematically:

$$Y = \frac{1}{c}\sum_{j}^{N} S_j Y_j,$$

where $c = \sum_{j}^{N} S_j$ is a normalization constant and S_j is the similarity score between the new subject and the j-th training subject.

16.3.3 Training, Validation, and Testing

SBML requires training to determine its parameters R, before application. Validation and testing ensure model performance:

1. Cross-validation: Divides data into training and validation sets to optimize parameters.
2. Bootstrapping: Randomly samples subsets for training and testing, useful for small datasets.

Overfitting occurs when a model fits training data too closely, leading to poor generalization. Regularization mitigates this by penalizing model complexity. For instance, with a mean squared error (MSE) loss function, ridge regularization minimizes:

$$L = \text{MSE} + \lambda |R|^2 ,$$

where $\lambda > 0$ controls the penalty for large scaling factors, balancing fit and generalization.

Learning is essentially the updating of the model parameters R so as to minimize the loss, using the gradient method algorithm with the training data. The gradient method makes the adjustment of R in the maximum slope direction (just as we go downhill following the steepest but shortest path). The scaling factor at iteration $t+1$ from iteration t is calculated using the formulation

$$R^{(t+1)} = R^{(t)} - \alpha \frac{\partial L}{\partial R}.$$

If we view gradient $\dfrac{\partial L}{\partial R}$ as the direction of walking, then the constant learning rate α (e.g., 0.125) determines the stride length. The learning rate should be small enough to have sufficient precision, but large enough (e.g., 0.25) for computational efficiency (Figure 16.2).

Now we know how to use SBML to resolve the Simpson paradox. We first record the responses in all patients and collect all potential relevant attributes, such as baseline disease severity, vital signs, gender, age, genomics, and other demographics. Training data are then used to determine the attribute-scaling factors for the attributes. The learned scaling factors govern

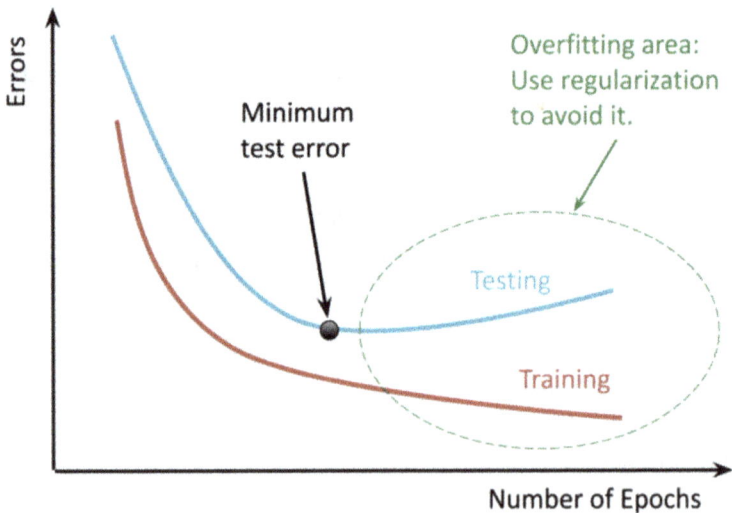

FIGURE 16.2
Comparing training and testing errors in machine learning.

the relative importance of each attribute in the similarity score. After the scaling factors are determined, the individual response is predicted using similarity-based weighting of the known responses. In short, to predict a patient's response, instead of basing the response (rate) in a predetermined category (e.g., all patients, male patients, or young female patients), SBML will weigh the responses of patients based on the similarity.

In summary, SBML operationalizes the similarity principle to address complex challenges in data analysis, such as Simpson's Paradox. By dynamically weighting similarities among subjects, SBML offers a robust framework for precision medicine and predictive modeling, ensuring that insights are both data-driven and personalized.

Application of SBML in cystic fibrosis, a rare, life-threatening disorder affecting multiple systems, can be found elsewhere (Hwang and Chang, 2022).

16.4 Summary: The Similarity Principle as a Foundation for Science and Decision-Making

This chapter delves into the similarity principle, underscoring its foundational role in scientific discovery, statistical inference, and decision-making. The principle asserts that similar entities often yield similar outcomes, forming the basis of analogical reasoning and prediction. This chapter seamlessly integrates critical thinking and creative analogies, showing how the similarity principle underpins scientific reasoning and modern machine learning, fostering innovative solutions across diverse fields.

Key Takeaways

1. Simpson's paradox illustrates how aggregated data can lead to different conclusions, which can be resolved through dynamic weighting using similarity-based reasoning.

2. By simplifying complex phenomena through similarity grouping, science and statistics make patterns observable while addressing the risks of oversimplification or hidden biases.

3. Examples like the traffic ticket paradox and life expectancy shift demonstrate how similarity-based reasoning shapes subjective priors and empirical evidence.

4. Everyday analogies, such as age-based grouping in schools or brand loyalty, demonstrate how the similarity principle influences decisions in education, marketing, and policy.

5. Highlights how similarity-based reasoning enables predictions, causal inferences, and statistical analysis, while the inherent subjectivity in defining "similarity" can be largely reduced using SBML.

Exercise

1. Explain the similarity principle in your own words. Why is it considered foundational to scientific reasoning and analogical thinking?

2. Describe Simpson's paradox. How does it illustrate the challenges of grouping data based on similarity? Provide a hypothetical example from a different field.

3. What critical thinking ideas and elements of creative analogies stand out to you in this chapter?

4. Analyze the paradox of life expectancy or the traffic ticket paradox from this chapter. Discuss how the choice of similarity group impacts conclusions in these scenarios.

5. Explain how SBML addresses issues like Simpson's paradox. How does the use of dynamic similarity weighting improve predictions and insights?

6. Create an analogy to explain the similarity principle to someone new to the concept. Use a scenario from daily life, such as choosing products or predicting behavior.

7. Propose an analogy or real-world example to show how similarity grouping influences decisions in fields like education, marketing, or healthcare.

8. Debate the ethical implications of using similarity-based predictions in sensitive areas like hiring, insurance, or law enforcement. How can these systems balance fairness and effectiveness?

9. Discuss whether the process of defining similarity is inherently subjective. How might this subjectivity affect scientific conclusions or machine learning outcomes?

10. Explore a specific application of SBML (e.g., predicting drug effectiveness or customer preferences). How does the approach differ from traditional statistical models?

17

iWordNet and Connotation of Understanding

Circular definitions are an inherent aspect of our understanding and communication.

There is no such thing as common knowledge; even if such knowledge exists, it is rooted in individual understanding and therefore varies from person to person.

17.1 Integrating Cognitive Science and Network Science: A New Frontier

Cognitive science, an interdisciplinary field dedicated to studying the mind and intelligence, integrates insights from diverse disciplines such as philosophy, psychology, artificial intelligence (AI), neuroscience, linguistics, and sociology. As the field evolves, AI-driven methodologies like statistical learning and network science are increasingly shaping research priorities. Among these, statistical learning has emerged as one of the most extensively explored phenomena in cognitive science. Complementing this is the growing application of network science, which has proven to be a powerful framework for advancing our understanding of cognitive processes and computational linguistics.

Network science has been successfully employed to address a wide array of challenges, including feature biases in early word learning, the structure of semantic concepts, grammatical relationships, spatial learning in human navigation, and the mapping of structural and functional brain connections linked to cognitive abilities. Additionally, network-based approaches have shown value in perceptual studies, such as texture and shape discrimination. Of all these domains, computational linguistics has perhaps embraced network science most robustly, showcasing its potential for transformative insights.

Despite its successes, the integration of network science into cognitive science remains incomplete. Current research primarily focuses on two key areas: (1) describing networks derived from sensory experiences and (2) investigating how the human brain engages with this sensory information. A critical gap persists in understanding how internal complex systems dynamics drive learning, how acquired knowledge shapes observable brain topology, and how micro- and macro-level neural processes collectively facilitate cognitive development.

DOI: 10.1201/9781003630081-17

An important parallel in this discussion is the study of artificial neural networks, particularly deep learning models inspired by biological neural networks. While this area has achieved remarkable advancements, its broader implications will be discussed in a separate chapter.

Inextricably linked to external networks—be they social relationships, transportation systems, or online platforms—our lives offer a fertile ground for network analysis. The study of social media networks, for example, has provided unprecedented insights into information transmission and human interaction. To fully harness the power of network science in studying the mind and cognitive development, it is imperative to conceptualize knowledge as a dynamic network of interconnected ideas, capturing both individual cognition and collective intelligence.

This perspective leads us to propose a novel approach: the development of individualized conceptual networks, termed iWordNets. These networks aim to map an individual's cognitive state or knowledge at a given time. By analyzing the topology of an iWordNet, researchers can uncover insights into cognitive development, individual differences in knowledge representation, and the unique trajectories shaping those differences. Such findings could inform personalized strategies for education, learning assessment, and clinical interventions.

The construction of an iWordNet involves a recursive process where a subject explains concepts sequentially, revealing the relationships and connotations between words. This approach aligns with the idea that understanding is inherently a mapping of words, each serving as a placeholder for meaning. Based on this framework, we hypothesize a correlation between the topology of an individual's iWordNet and their overall knowledge or IQ. Preliminary results from a pilot study support this hypothesis, suggesting a positive relationship between iWordNet complexity and cognitive capacity.

Building on these initial findings, we introduce the concept of the path of understanding (PoU)—a method for studying how individuals comprehend concepts, sentences, or paragraphs. By analyzing topological features (e.g., degree and centrality) along the PoU, we aim to quantify how new knowledge is acquired and internalized. The iWordNet framework thus offers a promising AI-driven, low-cost, and noninvasive tool for exploring the brain, mind, and cognitive processes, presenting exciting opportunities for future research in cognitive science and artificial intelligence.

17.2 The Connotation of Understanding

Knowledge is a structured collection of concepts, built from countless pieces of individual "understanding." Yet, we rarely pause to question the nature of understanding itself. What does it mean to "understand" a concept? Does each person hold a unique interpretation of a concept, and to what extent do these interpretations vary? Furthermore, what role does understanding play in communication and the exchange of ideas?

Knowledge can be seen in forms such as a library, a collection of books, a research paper, or the news—all repositories of concepts. A library, for instance, is a relatively straightforward, organized repository of knowledge. Similarly, social media networks are repositories, though more dynamic in nature. In a dictionary, words are defined by other words, which are further defined by still more words, forming an intricate web of connections. If we visualize these connections by linking related words with lines, we create a network of words—an external representation of knowledge independent of any one individual.

This approach mirrors how people explain concepts: A concept is explained using other concepts, which are in turn explained by yet more concepts, forming a recursive structure. By connecting a concept to the concepts directly involved in its explanation, we can construct a network of concepts (Figure 17.1). This network reflects the state of mind of the individual, offering a potential tool for studying their knowledge, learning processes, cognitive abilities, and more. Such a network serves as an external representation of a person's knowledge, which we call an **iWordNet** (short for "Individual Word Network"). The "i" emphasizes the personalized and individualistic nature of the network.

An iWordNet evolves over time, differing not only from person to person but also within the same individual as they learn and adapt. Studying iWordNets can shed light on cognitive development and help explain

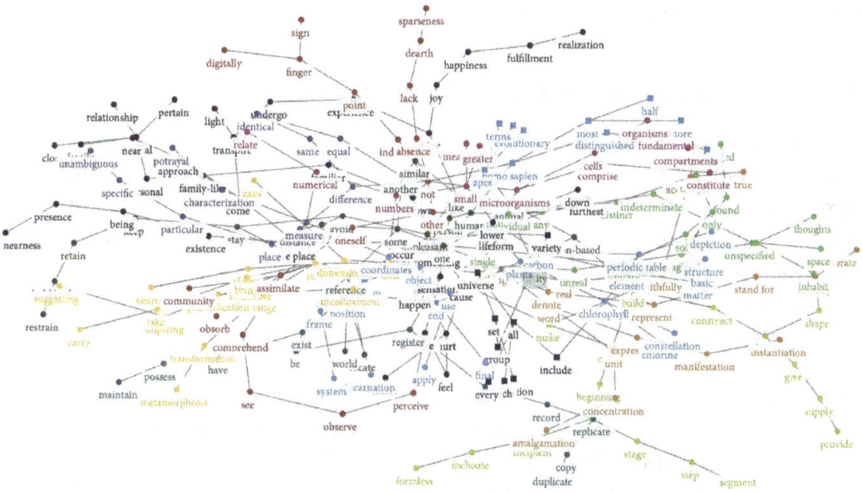

FIGURE 17.1
A partial iWordNet representation.

variations between individuals and groups. Furthermore, if an iWordNet is constructed for an entire organization, it can reflect the collective intelligence of that organization, providing insights into its behavior and even predicting its successes and failures.

Before delving into these applications, however, we must address the underlying statistical questions related to iWordNets. Specifically, we need to define "understanding" and "meaning" within this framework. This requires examining the connotations and structure of understanding.

17.2.1 The Nature of Understanding and the Paradox of Circular Definitions

A concept is always defined by other concepts, which themselves require definitions, creating an inevitable recursion. Given that individuals possess a finite set of concepts, this process results in an iWordNet where all definitions are either circular or terminate in concepts that cannot be further defined. Moreover, the meaning of each concept varies among individuals and shifts over time, influenced by experience and context.

In practice, understanding between individuals occurs when they establish a mapping—a shared perception of meaning. This exchange is often expressed in phrases like, "I know what you mean," or "I understand what you're saying." Importantly, this perceived understanding need not align with actual understanding. As long as each party believes they comprehend the other, communication appears successful.

In this framework, the meaning of a concept is not an intrinsic property but rather a mapping within an iWordNet, shaped by its topology. In this sense, meaning becomes fluid, subjective, and, paradoxically, almost meaningless— a phenomenon we refer to as the paradox of understanding.

17.2.2 Implications of Circular Definitions

Because definitions are circular, the specific concept at a given node in the iWordNet is insignificant and can be replaced with any symbol, such as a word from another language or a differently shaped node. Consequently, the topology of the iWordNet—the structure and connections within the network—becomes the defining characteristic of an individual's overall knowledge.

This perspective suggests that a person's level of knowledge, learning style, or cognitive capacity can be characterized by analyzing the topology of their iWordNet. The recursive nature of understanding, its reliance on circular definitions, and its contextual dependence open up new avenues for studying human cognition, learning processes, and even collective intelligence through the lens of network science.

17.3 Modeling Cognition with the iWordNet

An iWordNet is not merely a network of synonyms. Instead, it represents the recursive process of using concepts to explain other concepts, emphasizing the necessity of understanding each concept before understanding the one being explained.

A key distinction between iWordNets and WordNet® (Lexical Databases) or ConceptNet (Commonsense Knowledge Bases) lies in their focus and application. While the latter two are relatively static, exhaustive repositories of words, concepts, and their relationships—providing a broad understanding—they lack any information specific to individuals. In contrast, iWordNets are dynamic, individualized representations of knowledge, making them a valuable tool for studying cognitive development at a personal level.

17.3.1 Local and Global Topological Properties of iWordNets

The local topological properties of an iWordNet include features such as vertices (nodes, words, etc.), degrees, clusters, betweenness centrality, closeness centrality, eigenvalue centrality, PageRank, geodesic distance, and "top items." Meanwhile, global topological properties encompass metrics like the total number of vertices, edges, and components; the clustering coefficient; mean geodesic distance; graph density; and any derived variables. These properties collectively offer insights into the structure and complexity of an individual's conceptual network.

17.3.2 Statistical Modeling with iWordNets in Pilot Study

Cognitive, emotional, and personality traits can be modeled using the topological properties of iWordNets through regression and other statistical techniques. Both supervised and unsupervised statistical learning methods provide additional ways to extract meaningful patterns and relationships.

A pilot study explored the relationship between IQ and iWordNets (Figure 17.2), involving 20 subjects (ages 16–60, male and female). Participants completed an IQ test assessing short-term memory, analytical thinking, abstract problem-solving, mathematical ability, and spatial recognition. This test provided an overall IQ score along with left-brain (memory, logic, and numerical) and right-brain (perception, spatial relations, and creativity) components.

To construct iWordNets, participants were asked to define the word "smart." Keywords in their definitions, along with subsequent definitions of those key words, formed the vertices of their iWordNet. These words were

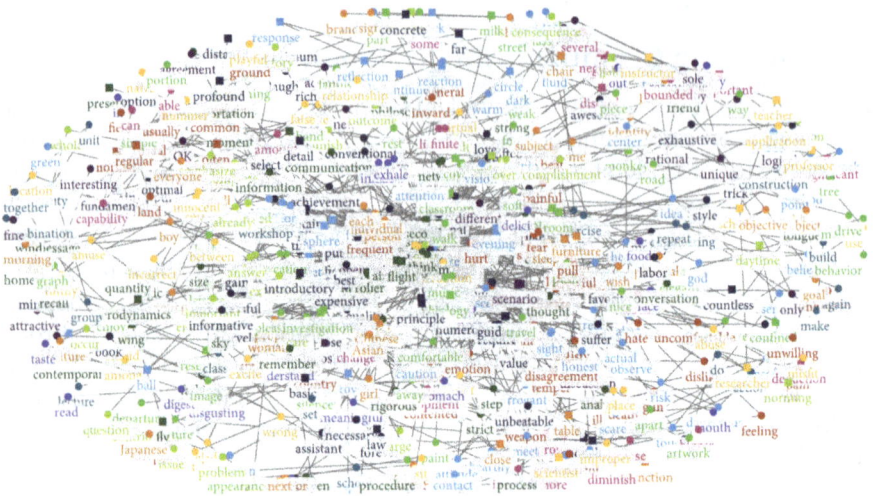

FIGURE 17.2
Visualization of the complete iWordNet model: a snapshot.

linked to create a personalized network of concepts. The iWordNets were then analyzed for topological properties such as degree, geodesic distance, and modularity, which were modeled against IQ scores using linear regression techniques. Tools used included NodeXL for network analysis, statistical software, and a computer with internet access for IQ testing.

The pilot study revealed that iWordNets exhibit a clustered structure, as identified by the Clauset–Newman–Moore clustering algorithm. This clustering reflects the intuitive observation that some words or concepts are more closely related than others. Modularity, a metric measuring the strength of a network's division into clusters, ranged from 0.53 to 0.76 for the iWordNets, indicating a strong clustering structure. Networks with high modularity feature dense connections within clusters but sparse connections between clusters.

The modularity results also revealed cognitive implications: Low modularity may indicate confusion among different concepts. High modularity may suggest limited cross-cluster connections, potentially hindering creativity by reducing the ability to make analogies across fields.

Other findings highlighted the small-world nature of iWordNets. Despite the number of vertices (words) ranging from 35 to 600, the mean geodesic distance (shortest path between two vertices) ranged only from 4.4 to 8.8, with diameters (maximum geodesic distance) between 12 and 18. This suggests that all words in the networks are closely related, a characteristic typical of small-world networks.

Interestingly, properties like diameter, mean geodesic distance, and modularity did not necessarily scale with the number of vertices, emphasizing the inherent efficiency and robustness of iWordNets. Similar to other small-world networks (e.g., electric grids, neural networks, and social networks), iWordNets are resilient to random deletion of vertices, meaning that forgetting a few words does not significantly disrupt the overall network structure. However, as with other small-world networks, iWordNets may be vulnerable to targeted attacks on central nodes, though no significant hubs were observed in this study (maximum degree ranged only from 10 to 15).

The study also observed: (1) Linear increases in the total number of unique edges and mean betweenness centrality as the number of vertices increased. (2) Exponential decreases in graph density and mean eigenvector centrality with growing network size. These findings reinforce the structured yet adaptable nature of iWordNets, where increasing complexity does not necessarily compromise network functionality.

17.3.3 Relationship between iWordNets and IQ

To examine how iWordNet topological properties correlate with IQ, nine variables were initially included in a linear regression model using backward elimination (p-value threshold of 0.1). The final model retained significant predictors:

- Mean degree ($p=0.0199$)
- Modularity ($p=0.0137$)
- Total number of edges ($p=0.0295$)
- Number of unique edges ($p=0.0093$)
- Graph density ($p=0.0798$)
- Stabilized density ($p=0.0612$)
- Stabilized centrality ($p=0.0589$)

The overall model achieved a p-value of 0.0002 and an R^2 value of 0.864, indicating that 86.4% of the variability in IQ scores was explained by the model. Figure 17.3 illustrates the predicted IQ scores (circles) versus observed values (squares), highlighting the model's accuracy.

In summary, this study demonstrates the potential of iWordNets as a novel tool for modeling cognitive and knowledge structures. By analyzing the topological properties of personalized conceptual networks, we can gain deeper insights into individual cognition, cognitive development, and even collective intelligence. The strong relationship between iWordNet properties and IQ suggests exciting opportunities for further research and application in areas such as education, psychology, and AI-driven cognitive modeling.

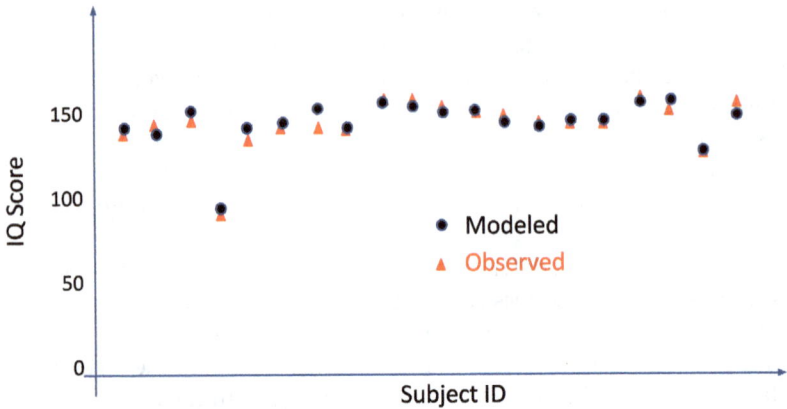

FIGURE 17.3
Comparing modeled and observed IQs in a pilot study.

17.4 The Path of Understanding

The global topology of an iWordNet offers a measure of an individual's overall knowledge and intelligence across different fields. However, it is insufficient for describing a person's understanding of specific concepts, such as individual words, phrases, or sentences. To capture this, we must examine the local topological properties of the iWordNet, as detailed below.

17.4.1 From Distributional Semantics to Individualized Meaning

In computational linguistics, distributional semantics associates the meaning of a word with its contextual usage—the words that appear near it in sentences. For instance, if the word "foot" were systematically replaced with "hand" in all contexts and vice versa, their meanings would effectively switch. This explains why different languages work equally well within their respective environments.

Building on this idea, iWordNets provide a more individualized approach. The meaning of a word, as understood through distributional semantics, can be viewed as an aggregate meaning. However, in an iWordNet, this meaning becomes personalized and dynamic, evolving alongside the individual's network. Consequently, the understanding of a word or concept shifts as the topology of the iWordNet changes. Yet, focusing solely on the topological properties of isolated words offers limited insight. Instead, we must study how strings of concepts are represented within an iWordNet, leading to the introduction of the PoU and its numerical characterization.

17.4.2 Defining the Path of Understanding

The PoU represents how an individual interprets a sequence of words (a phrase, sentence, or paragraph) within their iWordNet. This path follows the sequence of words in the text string and maps the associated topological properties (e.g., degree, centrality, and geodesic distance) of each word's node in the network. The resulting sequence of numerical vectors forms what we call a numerization string, which serves as a quantitative representation of the individual's understanding.

17.4.3 Illustration: Numerization String in Action

Consider the sentence:

"We can use a network approach to study the mind."

If an individual's iWordNet assigns degrees to the words as follows:

we: 5, **can:** 4, **use:** 7, **a:** 3, **network:** 2, **approach:** 9, **to:** 12, **study:** 6, **the:** 3, **mind:** 8

Then the numerization string for the sentence is:

```
{5, 4, 7, 3, 2, 9, 12, 6, 3, 8}
```

This string captures the person's understanding of the sentence based on the degree of each word's node in their iWordNet. Since different individuals have unique iWordNets, their numerization strings for the same sentence will differ, reflecting their distinct interpretations.

To further refine this representation, additional local properties, such as centrality, can be incorporated. For example, if the centrality values of the words are:

we: 0.001, **can:** 0.003, **use:** 0.0012, **a:** 0.0023, **network:** 0.0015, **approach:** 0.0032, **to:** 0.004, **study:** 0.0012, **the:** 0.003, **mind:** 0.0026

Then the numerization string becomes:

```
{(5, 0.001), (4, 0.003), (7, 0.0012), (3, 0.0023), (2, 0.0015),
(9, 0.0032), (12, 0.004), (6, 0.0012), (3, 0.003), (8, 0.0026)}
```

By incorporating multiple topological properties, the numerization string provides a more detailed characterization of an individual's understanding. Moreover, this representation can be extended to include three or more properties for even greater precision. Interestingly, numerization strings can also be visualized or converted into sound or music, creating new possibilities for interpreting understanding.

The numerization string can serve as the foundation for individualized statistical learning using tools like Bayesian inference, artificial neural networks, and deep learning models.

17.5 Acquiring New Knowledge

A person acquires new knowledge by expanding and adjusting their iWordNet, akin to applying smoothing techniques in control theory. For instance, when encountering an unfamiliar word, such as "romantic" in the sentence "I love romantic stories," the individual infers its meaning from the surrounding context. This inference relies on properties like geodesic distance (closeness) and degree (commonness) to make a reasonable guess.

For example, if potential substitutes for "romantic" are "enigmatic" and "humorous", probabilities from past experiences (e.g., 3/5 for "enigmatic" and 2/5 for "humorous") can guide the interpretation. In cases with multiple options, techniques like maximizing cosine similarity or employing statistical learning tools further refine understanding. Even if different PoUs produce identical cosine similarity scores, the associated probabilities help differentiate among potential meanings.

The paradox of knowledge acquisition: This process raises a fundamental question: if new concepts are understood through the existing ones, what constitutes the essence of new knowledge? This paradox highlights the iterative and evolving nature of understanding, where the acquisition of knowledge continually reshapes the iWordNet. Each new concept contributes to the network's topology, enhancing the individual's capacity for further learning and adaptation.

17.6 Summary: Understanding Through iWordNet and Cognitive Networks

This chapter introduces iWordNet, a personalized knowledge network that models the recursive nature of understanding through interconnected concepts. By framing knowledge as a dynamic network topology, the chapter highlights how meaning emerges from circular definitions and explores the PoU as a tool to quantify individual comprehension. Pilot studies demonstrate correlations between iWordNet properties and cognitive measures like IQ, emphasizing the framework's potential applications in education, AI, and cognitive science.

Key Takeaways

1. Cognition as a network: This chapter highlights how understanding is inherently a mapping of interconnected concepts within recursive networks, evolving as knowledge grows. The paradox of understanding, where circular definitions sustain meaning, challenges traditional views of fixed knowledge structures.

2. Quantifying understanding: Tools like iWordNet topology and PoU numerization strings enable the statistical modeling and analysis of cognitive processes, providing insights into individual learning and intelligence.

3. Visualizing knowledge: iWordNet is analogized to a social network, where concepts act as nodes and their relationships as edges, underscoring the personalized and interconnected nature of knowledge.

4. Dynamic representation: iWordNet offers a dynamic, evolving model of cognition that facilitates a deeper understanding of individual knowledge structures and their development over time.

5. Applications in cognitive science and beyond: By linking iWordNet properties to cognitive measures, the framework provides tools for exploring intelligence, learning, and collective knowledge, with diverse applications ranging from diagnosing cognitive impairments to enhancing personalized education and advancing AI systems.

6. Advancing AI: The principles of iWordNet could inspire the development of AI systems that better mimic human understanding by modeling networked relationships between interconnected concepts.

Exercise

1. Explain the paradox of understanding in your own words. Why are circular definitions unavoidable in knowledge representation?

2. Define iWordNet and describe how it differs from traditional static knowledge networks like WordNet®. How does it capture individual cognition?

3. What critical thinking ideas and elements of creative analogies stand out to you in this chapter?

4. Explain how the PoU numerization string works. Provide an example using a short sentence and outline how its representation might differ between two individuals.

5. Discuss how the topological properties of iWordNets (e.g., modularity and graph density) relate to cognitive measures like IQ. What insights can these correlations provide about individual learning or intelligence?

6. Create an analogy to explain the iWordNet framework to someone new to the concept. Use a real-world system like social networks or transportation maps to illustrate your explanation.

7. Propose an analogy to describe how an iWordNet evolves as new knowledge is acquired. Highlight the parallels between this process and real-life systems that adapt over time.

8. Debate whether the subjective nature of understanding (as modeled by iWordNets) challenges the concept of universal knowledge. How might this impact education or AI development?

9. Discuss the potential of iWordNets for advancing artificial intelligence and personalized education. What ethical or practical challenges might arise from these applications?

10. Design a simple experiment to construct an iWordNet for a concept like "learning." Identify the steps and tools you would use and the type of insights you might gain from its analysis.

11. Explore how iWordNets could model collective intelligence in organizations or societies. What might the topological differences between individual and collective networks reveal?

18

Analogy in Artificial Intelligence

Artificial intelligence uncovers the secrets of human intelligence through the power of analogy.

18.1 Innovative Ideas in AI Architecture Evolution

18.1.1 Classification of Machine Learning

Machine learning methods can be classified into five categories: supervised, unsupervised, reinforcement, swarm intelligence, and evolutionary learning (Figure 18.1).

Supervised learning is trained on labeled data, where both inputs and corresponding outputs are provided. The goal is to learn a mapping function from inputs to outputs to make predictions on unseen data.

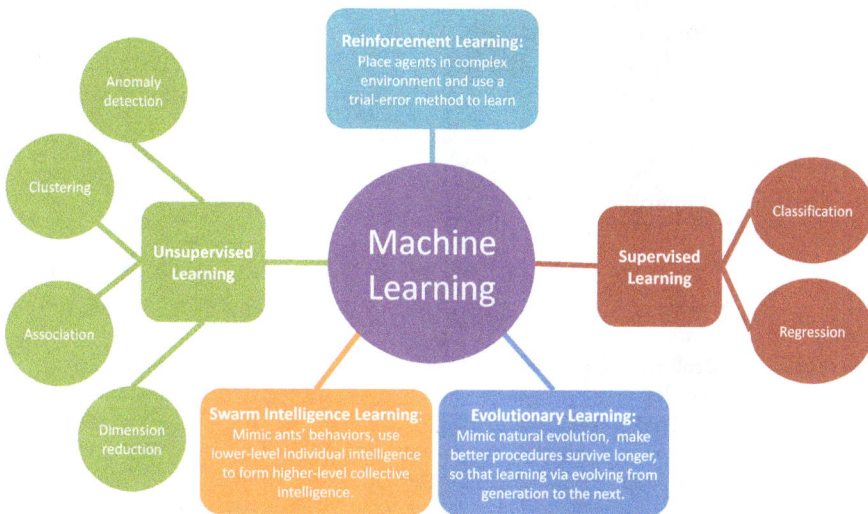

FIGURE 18.1
Classification of machine learning methods.

DOI: 10.1201/9781003630081-18

Unsupervised learning is trained on unlabeled data, and the task is to identify patterns or structures in the data. The goal is to explore the data's intrinsic properties and group or simplify it.

Reinforcement learning learns by interacting with an environment and receiving feedback in the form of rewards or penalties. The goal is to develop a policy that maximizes cumulative rewards over time.

Swarm intelligence learning, inspired by the collective behavior of decentralized, self-organized systems (e.g., ants, bees, and birds). The goal is to solve optimization or search problems using simple agents that communicate locally.

Evolutionary learning, inspired by biological evolution, where the model uses mechanisms like selection, mutation, and crossover to evolve solutions. The goal is to find optimal solutions to problems by iteratively improving a population of candidate solutions.

Table 18.1 is a comparative summary of different methods. These methods represent different approaches to solving problems in machine learning. While supervised and unsupervised learning are foundational, reinforcement, swarm intelligence, and evolutionary learning are specialized paradigms with unique applications. Understanding their strengths and limitations allows researchers and practitioners to choose the right method for a given problem.

18.1.2 Genetic Programming (GP)

GP is an evolutionary algorithm inspired by the principles of natural selection, designed to evolve computer programs, mathematical models, or engineering solutions. It is particularly valuable for problems where explicit algorithms or mathematical equations are unknown, making it highly effective in addressing complex, non-linear challenges across various domains.

TABLE 18.1

Comparison and Interplay

Aspect	Supervised	Unsupervised	Reinforcement	Swarm Intelligence	Evolutionary
Data requirement	Labeled data	Unlabeled data	Interaction data	Population-based data	Population-based data
Feedback signal	Explicit labels	No feedback	Rewards/ Penalties	Emergent behavior	Fitness function
Optimization	Gradient-based	Pattern discovery	Policy optimization	Collective optimization	Evolutionary search
Adaptability	Static tasks	Static patterns	Dynamic environments	Flexible dynamics	Flexible dynamics
Computational cost	Moderate	Low to moderate	High	Low to moderate	High

In mathematics, GP is widely applied in symbolic regression, where it uncovers the underlying mathematical relationships between variables without assuming a predefined model structure. For example, given raw data from an unknown physical process, GP can derive an equation such as $f(x) = a \cdot \sin(bx) + cx^2 + d$, identifying both the form and coefficients of the equation. Additionally, GP is used to approximate solutions to differential and integral equations, providing a practical approach for systems that lack analytical solutions. It is also effective for function optimization, particularly in non-linear, multi-objective problems where traditional optimization techniques fall short.

In engineering, GP plays a critical role in design optimization. It can evolve designs for structures, circuits, and mechanical components to enhance performance while minimizing material use and cost. For instance, GP has been used to optimize aerodynamic wing shapes, reducing drag and improving efficiency. In control systems, GP develops adaptive algorithms for robotics, industrial automation, and process controls, enabling systems to adjust dynamically to real-world conditions. GP is also applied in fault detection and diagnostics, where it identifies anomalies and predicts failures in machinery, electrical grids, and production processes. In signal processing, GP evolves filters and waveforms for noise reduction, signal reconstruction, and data compression, improving the quality and efficiency of data handling. GP is also implemented in the HAI architectures in Chapter 19.

The key advantage of GP lies in its ability to solve problems without requiring prior knowledge of the system's model. It automates the discovery of solutions, making it flexible for a wide range of applications. However, GP comes with challenges, including its high computational demands and the tendency to produce overly complex models that may be difficult to interpret. Additionally, careful tuning of parameters such as mutation rates and population size is essential to ensure convergence to optimal solutions.

18.1.3 Swarm Intelligence Learning

Systems in which organized behavior arises without a centralized controller or leader are often called self-organized systems, while the intelligence possessed by the system is called swarm intelligence (SI) or collective intelligence. Examples of swarm intelligence in nature include ant colonies, bird flocking, hawks hunting, animal herding, bacterial growth, fish schooling, and microbial intelligence.

The SI characteristics of a human network integrate two correlated perspectives on human behavior: cognitive space and social space. In SI, we see the evolution of collective ideas, not the evolution of people who hold ideas. Evolutionary processes have costs: redundancy and futile exploration. Such processes are necessary to be adaptive and creative. The system parameters of SI determine the balance of exploration and exploitation.

An SI algorithm comprises a population of individuals that interact with one another according to simple rules in order to solve problems. Individuals in an SI system have mathematical intelligence (logical thought) and social intelligence (a common social mind). Social interaction thus provides a powerful problem-solving algorithm in SI.

An ant is simple and (arguably) dumb, while a colony of ants is complex and intelligent. Likewise, neurons are simple but brains are as complex as a swarm. Competition and collaboration among cells lead to human intelligence; competition and collaboration among humans form a social intelligence, or what we might call the global brain. Nevertheless, such intelligence is based on a human viewpoint, and thus it lies within the limits of human intelligence. Views of such intelligence held by other creatures with a different level of intelligence could be completely different!

SI has some similarities to ensemble intelligence (EI), but they are different, in that each individual in SI has no intelligence, while each individual in EI is usually an expert. SI is necessarily the consequence of collective dumbness, a result of collaboration, while EI can be just the best opinion among or average opinion of the experts without any collaboration at all. Another difference is that SI requires a larger population of the same type of individuals, while a "population" in EI populations usually only involves a small set of individuals obtained by different methods.

SI is also different from reinforcement learning. In reinforcement learning, an individual can improve his level of intelligence over time since, in the learning process, adaptations occur. In contrast, SI is a collective intelligence from all individuals. It is a global or macro behavior of a system. In complex systems, there are a huge number of individual components, each with relatively simple rules of behavior that never change. However, in reinforcement learning, there are not necessarily a large number of individuals; in fact there can just be one individual with built-in complex algorithms or adaptation rules.

Particle swarm optimization (PSO), one of the bio-inspired algorithms, is a stylized representation of the movement of organisms in a bird flock to search for a problem's optimal solution. PSO is a metaheuristic, as it makes few or no assumptions about the problem being optimized and can search very large spaces of candidate solutions. PSO does not use the gradient of the problem being optimized. However, PSO does not guarantee an optimal solution is ever found. Each particle's movement is influenced by its local best known position, but is also guided toward the best known positions in the search-space. This is expected to move the swarm toward the best solutions.

A basic variant of the PSO algorithm works by having a population (called a swarm) of candidate solutions (called particles). These particles are moved around in the search-space in three possible directions: (1) the personal best direction, (2) the swarm's best-known direction, and (3) its current direction. We can randomly decide on one of the three directions or take a weighted

Goal: A team of Three to Find a Mountain's Summit

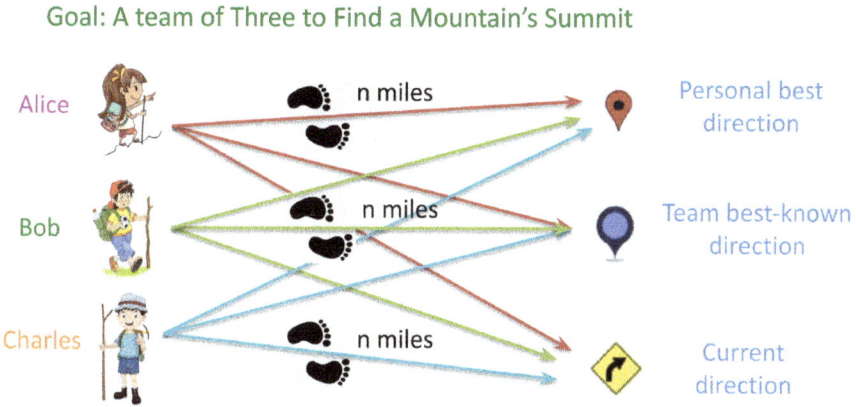

Strategy: Everyone walks n miles in the current direction, then n miles in the personal best direction, and then n miles in the team-best direction. Repeat the process

FIGURE 18.2
A three-mountaineer team demonstrating PSO.

three-directional vector. The process is repeated, and by doing so it is hoped, but not guaranteed, that a satisfactory solution will eventually be discovered. Figure 18.2 illustrates how a three-person team climbs to a mountaintop using the PSO search strategy.

18.1.4 Feedforward Networks

Artificial neural networks (ANNs) are computing systems inspired by the biological neural networks in animal brains. ANNs take input data and output desired outcomes after training. The learning in an ANN refers to its ability of outputting outcomes that, through training, are closer and closer to the right answer over time. The adjustments of weights in an ANN are what make the ANN learn.

An ANN model (Figure 18.3) includes the input layer, one or more hidden layers, and the output layer. Each layer contains input and output nodes, weights, and activation functions. Deep learning ANN architectures include (1) *feedforward neural networks* (FNNs) for general classification and regression, (2) *convolution neural networks* (CNNs) for image recognition, (3) *recurrent neural networks* (RNNs) for speech recognition and natural language processing (NLP), and (4) *deep belief networks* (DBNs) for disease diagnosis and prognosis. Two other popular neural networks are *generative adversarial networks* (GANs) for classification problems and *autoassociative networks* (*Autoencoders*) for dimension reduction. Although an autoencoder will result in a dimension reduction in an unsupervised manner, the training process is supervised learning.

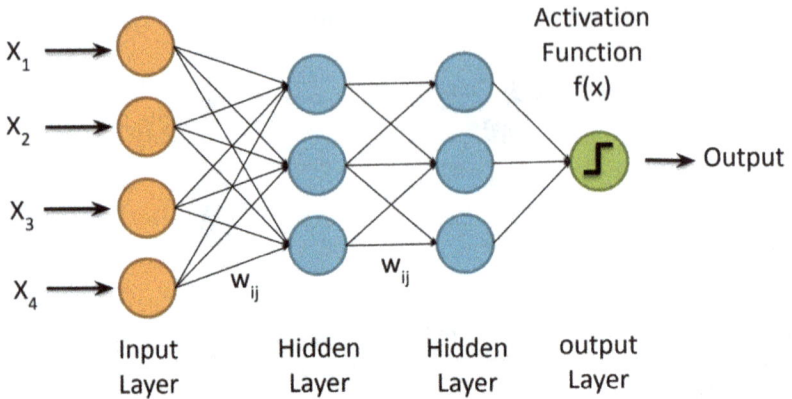

FIGURE 18.3
FNN structure.

An FNN, also known as a multilayer perceptron, has input and output layers and hidden layers in between. At each layer (except the input layer), an activation function f is applied to the weighted sum of input data from the previous layer. The resulting outputs at each layer serve the input data for the next layer. For example, the output Y_i for the ith node at the first layer is a function of input data x_j,

$$Y_i = f\left(\sum_j w_{ij} x_j\right).$$

The outputs Y_i will serve as the input for the next layer and so on. Different activation functions f can be used at different layers in an ANN, such as the *rectifier* (ReLU), *sigmoid*, and *tanh* functions used to mimic a biological mechanism.

The key learning algorithm ANNs is the so-called backpropagation algorithms as decribed below.

The numbers of layers and nodes are usually fixed; the only things that can change are the weights in the network. The question is how to convert a person's way of learning into a set of rules for changing the weights so that the network outputs the right answer or appropriate response more often. In practice, the weight modifications are through training using the gradient method, more precisely, a backpropagation algorithm (BPA).

BPAs (Bryson and Ho, 1975) make deep learning ANNs computationally possible. In fact, a BPA for multilayer ANNs was an important precursor contribution to the success of deep learning in the 2010s, once big data became available and computing power was sufficiently advanced to accommodate the training of large networks. There are several AI software packages available in R for building ANNs, including *keras* and *kerasR*.

18.1.5 Convolutional Neural Networks

18.1.5.1 *Ideas behind CNNs*

A CNN is actually a class of deep neural networks, mainly applied to image analysis. From an analogy perspective, convolution layers in CNNs can be understood as mimicking natural processes of perception and filtering, particularly how humans and other organisms visually process their surroundings. Here is a detailed analogy-based exploration.

A CNN architecture consists of many layers (Figure 18.4), each one playing a different role. (1) The *input layer* takes the input from the source images or objects and converts it to data or numbers. (2) A *convolution layer* identifies certain features of the images by inspecting the image's pieces and outputting a value dependent on the filter used. A filter is a powerful tool that makes it possible to discover a feature contained in the source images. To identify different elemental features, we use filters at different convolution layers. (3) An *activation layer* decides whether the neuron fires ("spikes") for the current inputs. (4) A *pooling layer* converts the original higher resolution images to lower resolution images, in order to reduce the size of the images. (5) Although some weights connecting layers can be removed (dropped) to reduce the dimension for computational efficiency, the fully connected layers (*dense layers*) take the high-level filtered images and translate them into votes in classifying the source images.

18.1.5.2 *Convolution Layers*

The main idea of a CNN is seen at the convolution layers, where different filters are used. Each filter is used to identify or filter out particular features or image elements such as eyes, noses, lines, etc., just as when we search for particular objects from a complex picture.

The term convolution is from mathematics (calculus). It corresponds to an image inspection process through a small moving filter. Let us look at how convolutions work in CNN using Figure 18.5.

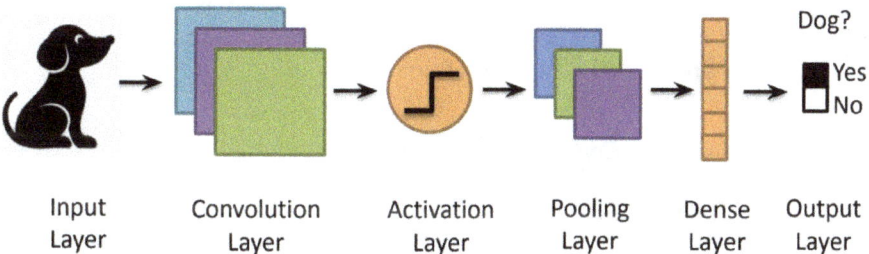

FIGURE 18.4
Deep learning architecture: a sketch of CNN design.

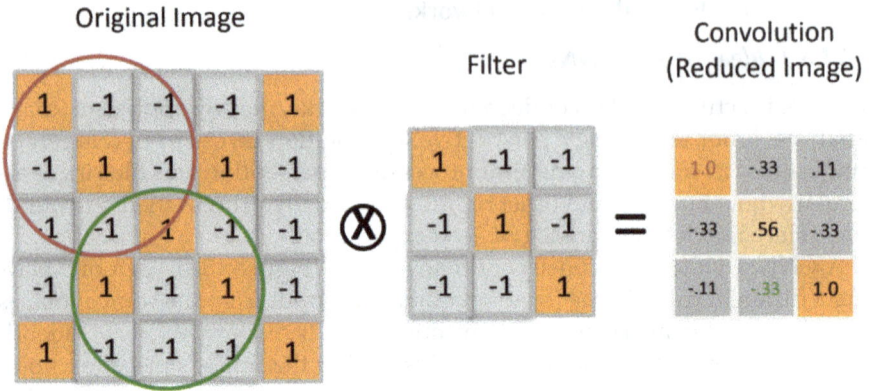

FIGURE 18.5
Demonstration of convolution leading to image reduction.

Taking image X as an example, we code a value of 1 for the pixels where X is located and a value of –1 for all other places. The filter with a backslash is also coded using 1 and –1 implementing the same rule.

To filter the image, we place a filter over the image, starting from the left upper corner, do the calculation (filtering) and then move to the next position by a stride (one or more pixels to the left or down) and perform filtering again. We continue until the filter covers all possible positions. To calculate the convolution at a position, we simply (1) put the filter on top of the coded image, (2) count the numbers of matches (pixels with the same code, –1 or 1) and mismatches, and (3) compute the proportion of net matches:

$$\text{Convolution} = (\text{\# of matches} - \text{\# of mismatches}) / (\text{size of filter in pixels}).$$

Obviously, filtering will result in a shrunken image (fewer pixels), but, more importantly, the filtered image does show larger values on the diagonal of the filtered image, indicating that the original image has a backslash we were searching for using the filter. If there were no backslash in the original picture as the filter is looking for, the resulting picture would not have a backslash that is represented by higher values.

18.1.5.3 Summary of CNNs (1989)

Logic Development

- CNNs introduced hierarchical feature extraction for grid-like data (e.g., images) using convolutional filters.
- Inspired by the visual cortex, they added pooling layers to reduce spatial dimensions and fully connected layers for classification.

Key Challenges Resolved

- Overcame the manual feature engineering required for computer vision tasks.
- Reduced the parameter space by sharing weights in convolutional layers.

Limitations

- Ineffective for sequential data due to fixed input size and lack of temporal relationships.

Impact

- Revolutionized computer vision, enabling applications like image recognition and object detection.

18.1.6 Recurrent Neural Networks

Location invariance and local compositionality are two key ideas behind CNNs that do not always bear fruit. They make sense for computer vision applications but not for NLP or time-series events. The location where a word lies in the whole sentence is critical to the meaning of the sentence. Words that are not close to one another in a sentence may be more connected in terms of meaning, which is quite contrary to pixels in a specific region of an image that may be a part of a certain object. Therefore, it makes sense to look for a neural network that reflects the sequence of the tokens, whether they are words, events, or something else with a temporal axis. One such network is the RNN, which can have memories of its previous states.

The idea of the RNN came from the work of Ronald William and his colleagues in 1986. An RNN is a class of ANNs for modeling temporal dynamic behavior. Unlike FNNs, RNNs can use their internal state as memory to process sequences of inputs. In other words, they often reuse the output or hidden outputs (internal states) as input again, hence their name. RNNs are useful for tasks such as unsegmented, connected handwriting recognition and speech recognition. They have also been implemented for stock market prediction, sequence generation, test generation, voice recognition, image captioning, poem-writing (after being trained in Shakespeare's poetry), reading handwriting from left to right, and creating music.

A challenging issue with RNNs is the vanishing gradient problem when the RNN involves many layers. Traditional activation functions such as the hyperbolic tangent function have gradients in the range (0, 1), and backpropagation computes gradients by the chain rule. This has the effect of multiplying n of these small numbers to compute gradients of the "front" layers in an n-layer network, leading to the gradient (error signal) decreasing exponentially with n while the front layers train very slowly. The vanishing gradient will effectively prevent the weight from changing its value and can even completely stop the neural network from further training. A solution is to use a long chain of short-term memory units, called long short-term memory units (LSTMs) proposed by Hochreiter and Schmidhuber (1997).

18.1.6.1 Summary of LSTM Networks (1997)

Logic Development

- LSTMs addressed the vanishing gradient problem in RNNs by introducing gated mechanisms: forget, input, and output gates.
- These gates allowed the network to selectively remember or forget information over long sequences.

Key Challenges Resolved

- Solved the inability of RNNs to learn long-term dependencies in sequences.
- Enabled tasks like language modeling and time-series prediction.

Limitations

- Computationally expensive due to sequential processing.
- Inefficient for very long sequences compared to attention mechanisms.

Impact

- Made sequence modeling viable for speech recognition, machine translation, and other temporal tasks.

18.1.7 Generative Adversarial Networks

GANs are deep neural net architectures composed of two nets, pitting one adversarially against the other. GANs can be viewed as the combination of a counterfeiter and a policeman, where the counterfeiter is learning to pass false notes, and the cop is learning to detect them. Both are dynamic in the zero-sum game, and each side comes to learn the other's methods in a constant escalation. As the discriminator changes its behavior, so does the generator, and vice versa. Their losses push against each other.

18.1.7.1 Summary of GANs (2014)

Logic Development

- GANs introduced a framework where two networks, a generator and a discriminator, compete: the generator creates data samples, and the discriminator evaluates their authenticity.
- The generator improves by fooling the discriminator.

Key Challenges Resolved

- Enabled high-quality synthetic data generation for images, videos, and more.
- Bypassed the need for explicit likelihood estimation in generative tasks.

Limitations

- Training instability due to mode collapse (generator producing limited diversity).
- Sensitive to hyperparameter tuning.

Impact

- Revolutionized generative tasks in art, media, and data augmentation.

18.1.8 Autoassociative Networks

An autoassociative network (autoencoder) is a type of ANN used to learn efficient data coding in an unsupervised manner. Autoencoders encode input data as vectors. They create a hidden, or compressed, representation of the raw data (Figure 18.6). Such networks are useful in dimensionality reduction; that is, the vector serving as a hidden representation compresses the raw data into a smaller number of salient dimensions. Autoencoders can be paired with a so-called decoder, which allows one to reconstruct input data based on its hidden representation. Autoencoders are especially useful for dimension reduction, but the training method used is supervised learning, since the correct answer is known for each input. The training goal is to minimize the error between the output and the input.

An autoencoder learns to compress data from the input layer into a short code and then decompress that code into something that closely matches the original data. A simple autoassociative network can be a multiple-layer perceptron, where the output is identical to the input and the middle hidden layer is smaller. This means we can use the compressed middle layer to generate the original image.

An autoencoder has a hidden layer that is smaller than the input layer, while the output is always the same (approximately) as the inputs. After extensive training, the smaller input at a hidden layer can securely generate (nearly) the same input image. Therefore, autoencoders can be used for compressing images or data.

18.2 Generative Pretrained Transformer

The introduction of transformers in 2017 revolutionized the field of NLP and beyond. By replacing recurrence with a self-attention mechanism, transformers enabled models to capture global dependencies in a sequence without

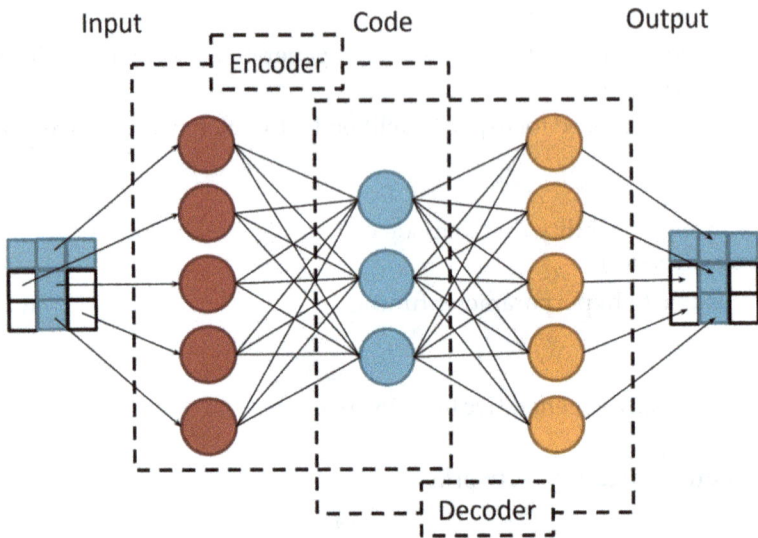

FIGURE 18.6
Autoassociative network applied to data compression.

requiring sequential computation. This was achieved through innovations like scaled dot-product attention and positional encodings, which preserved the order of tokens in a sequence. These advancements addressed the inefficiencies of LSTMs by allowing parallel processing, enabling models to handle much longer sequences efficiently while enhancing contextual understanding by attending to all tokens simultaneously. Despite their computational expense for very long sequences due to quadratic complexity in attention, transformers became foundational for state-of-the-art models in NLP and vision, including BERT, generative pre-trained transformer (GPT), and vision transformers.

Building on the transformer architecture, GPTs emerged between 2018 and 2023 as a major leap forward. GPT models employ a two-stage process: pre-training on extensive corpora using unsupervised learning, followed by fine-tuning for specific tasks. These models generate text autoregressively by predicting the next token in a sequence, showcasing the power of transfer learning in language tasks. This approach improved scalability and adaptability for a wide range of NLP applications with minimal task-specific training. However, GPT models also face challenges, such as high computational and memory requirements during training and the potential to produce inaccurate or biased outputs due to training data. Despite these limitations, GPT has set new benchmarks in NLP, excelling in tasks such as text generation, summarization, and conversational AI. Together, transformers and GPT represent a paradigm shift in AI, enabling unprecedented advancements in understanding and generating human language.

18.2.1 Analogies in GPT

From an analogy perspective, GPT's architecture reflects and builds upon fundamental concepts inspired by human cognition and language understanding. Here is how GPT can be analyzed through the lens of analogy:

1. Self-attention and global context: A conversation participant who maintains awareness of the entire discussion. GPT's self-attention mechanism mirrors how humans process language by considering the context of every word or phrase within a larger conversation. It "attends" to all parts of a sentence or passage to determine their relationships, much like how people relate current words or ideas to the broader narrative.

2. Positional encodings and temporal understanding: A storyteller remembering the sequence of events. The inclusion of positional encodings parallels a storyteller's ability to remember the sequence in which events occur. GPT integrates positional information as trainable parameters, akin to dynamically learning the importance of sequence and timing in language.

3. Autoregressive generation: Predicting the next word in a sentence based on prior context. GPT's autoregressive nature reflects how humans construct sentences, one word at a time, by leveraging prior context. This sequential approach mimics the natural flow of thought and expression, allowing for coherent text generation.

4. Transfer learning: Gaining expertise from diverse experiences before specializing. GPT is pre-trained on a massive corpus of text (general learning) and fine-tuned for specific tasks (specialized learning). This process is analogous to a person developing general knowledge through broad reading and then applying it to a specific domain or task.

5. Scaled dot-product attention: Assigning importance to relevant ideas in a discussion. The scaled dot-product attention mechanism in GPT mirrors how people focus on the most relevant ideas or keywords during communication, filtering out less important information while emphasizing critical points.

6. Layered depth and hierarchical learning: Organizing thoughts into layers of abstraction. GPT processes language through stacked transformer layers, analogous to how humans organize thoughts hierarchically—from raw sensory inputs to high-level abstract reasoning.

7. Emergent behavior: Creativity emerging from combining learned patterns. GPT's ability to generate novel ideas and responses by recombining learned patterns mirrors human creativity, which often stems from drawing analogies between previously unrelated concepts.

8. Handling ambiguity: Contextual interpretation in conversation. GPT resolves ambiguity in language by weighting multiple possibilities and refining its understanding based on context, much like humans resolve ambiguities through discussion and context.

To summarize, by drawing analogies with human cognition and language processing, GPT integrates core principles of how humans understand and generate language. These analogical insights guided the design of its architecture, enabling it to achieve state-of-the-art performance in tasks requiring comprehension, reasoning, and creativity. Analyzing GPT from an analogy perspective not only deepens our understanding of its design but also offers a roadmap for future advancements in AI.

18.2.2 Architectures of GPT

The mathematical model for GPT is built on the transformer architecture. It leverages a self-attention mechanism and is designed for sequence modeling and language generation tasks.

18.2.2.1 Input Representation

Each input text is tokenized into a sequence of tokens $x = (x_1, x_2, \ldots, x_n)$, where n is the sequence length.
   Tokens are converted into vectors via an embedding layer:

$$E_x = (e_{x_1}, e_{x_2}, \ldots, e_{x_n}),$$

where $e_{x_i} \in R^d$ is the embedding of token x_i, and d is the embedding dimension.
   Positional encodings $P = (p_1, p_2, \ldots, p_n)$ are added to include order information:

$$H_0 = E_x + P.$$

18.2.2.2 Transformer Block

GPT uses stacked transformer blocks. Each block consists of two main components:

1. Self-attention mechanism.
2. FNN.

The self-attention mechanism computes a weighted sum of token embeddings for each token, using the scaled dot-product attention:

$$\text{Attention}(Q, K, V) = \text{softmax}\left(\frac{QK^T}{\sqrt{d_k}}\right)V,$$

where Query matrix $Q = H_l W_Q$, Key matrix $K = H_l W_K$, Value matrix $V = H_l W_V$, and W_Q, W_Q, and W_Q are learnable weight matrices.

After the self-attention layer, a pointwise FNN is applied:

$$FFN(H_l) = ReLU(H_lW_1 + b_1)W_2 + b_2,$$

where W_1, W_2, and b_1, b_2 are learnable parameters.

Each sub-layer is followed by residual connections and layer normalization:

$$H_{l+1} = LayerNorm(H_l + selfAttention(H_l)),$$

$$H_{l+1} = LayerNorm(H_l + FFN(H_l)).$$

Here H_l is the input to the self-attention sub-layer at layer l, and the residual connection adds H_l back to the result of the self-attention operation before normalization.

The first equation represents the output of the self-attention sub-layer, before passing through the feedforward sub-layer and The second equation represents the output after the feedforward sub-layer. These equations succinctly describe the structure of each layer in the transformer architecture.

18.2.2.3 Output Layer

The final transformer block outputs a representation H_L, where L is the number of layers.

To predict the next token x_{n+1}, a linear layer and softmax are applied:

$$P(x_{n+1} \mid x_1, x_2, \ldots, x_n) = softmax(H_L W_o + b_o),$$

where W_o and b_o are the output weights and biases.

18.2.2.4 Training Objective

GPT is trained to maximize the likelihood of the next token:

$$L(\theta) = -\sum_{i=1}^{N} \log P_\theta(x_{i+1} \mid x_1, x_2, \ldots, x_i),$$

where θ represents all trainable parameters in the model.

18.3 AI Diffusion Model

AI diffusion models (Figure 18.7) are advanced generative models that transform random noise into structured, realistic data through a probabilistic two-phase process. In the forward phase (destruction), data is incrementally

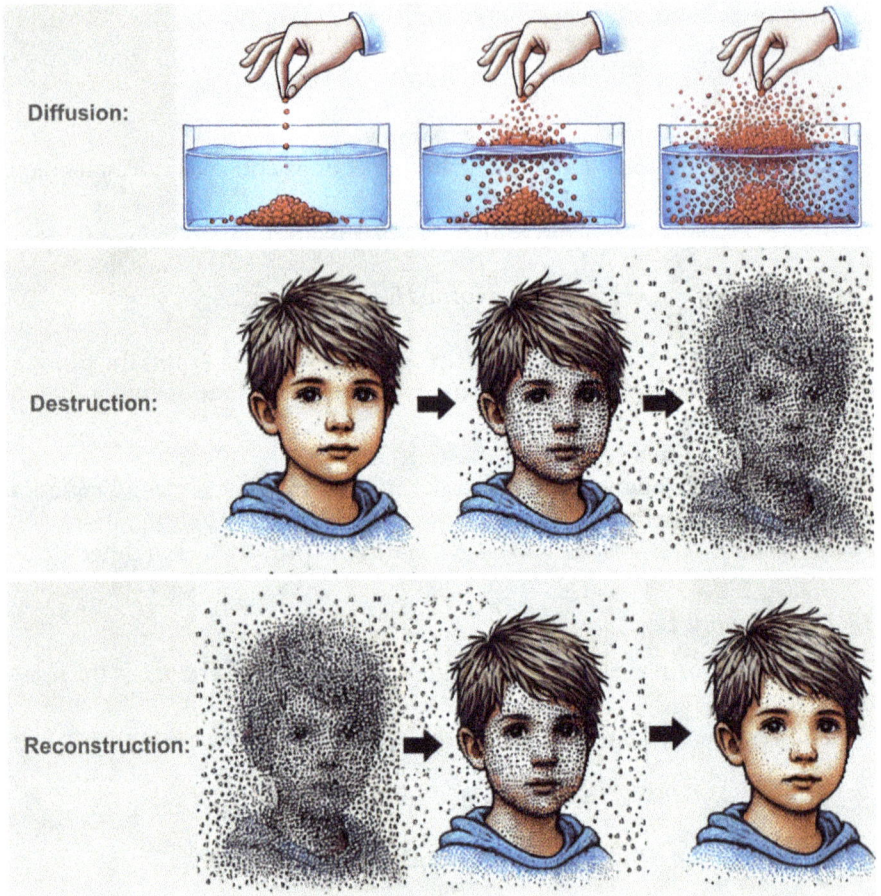

FIGURE 18.7
AI diffusion model inspired by particle diffusion.

corrupted with noise, and in the reverse phase (reconstruction), the model iteratively denoises to reconstruct high-quality outputs. Known for their versatility, diffusion models are used in applications such as image generation (e.g., DALL·E 2 and Stable Diffusion), audio synthesis, video creation, and even drug discovery. They offer high fidelity, stability in training, and applicability across multiple data modalities. However, they can be computationally intensive and complex to interpret. Recent developments focus on improving efficiency, combining diffusion models with other generative techniques, and democratizing access through open-source frameworks, solidifying their role in advancing generative AI.

18.3.1 Analogies in AI Diffusion Model Architectures

From an analogy perspective, the architecture of AI diffusion models can be likened to natural and human processes involving gradual refinement, noise reduction, and pattern recognition. Here's a detailed analogy-based analysis of their architecture:

1. Forward and reverse processes as resembling sculpture creation from a rough block: The **forward process**, where structured data is progressively corrupted with noise, is analogous to starting with a well-defined block of material and incrementally chiseling away until it becomes an unrecognizable lump. The **reverse process**, where noise is gradually removed to reconstruct meaningful data, mirrors the sculptor's deliberate efforts to shape the block into a detailed and recognizable figure. The iterative nature of this process ensures precision and alignment with the intended result.

2. Iterative refinement mimicking learning from mistakes: The step-by-step denoising in diffusion models resembles how humans learn iteratively, refining their understanding or skills by correcting small errors at each stage. Each step builds on the previous one, improving accuracy and quality over time.

3. Probabilistic modeling as an analogy for forecasting with uncertainty: Diffusion models rely on probabilistic frameworks to model noise addition and removal, much like how humans make predictions with uncertainties in complex scenarios (e.g., weather forecasting or stock market analysis). These probabilities help guide the process toward plausible outcomes.

4. Noise as a starting point: Creativity emerging from chaos: Starting with random noise reflects how creative processes often begin with vague or chaotic ideas. Over time, patterns and coherence emerge as the process refines and organizes the chaos into meaningful outputs.

5. Global and local context simulating solving a puzzle with all pieces in view: During the reverse process, the model attends to both local details (specific pieces of the puzzle) and global context (how the pieces fit together). This mirrors how humans solve puzzles or synthesize ideas by constantly switching focus between the small details and the bigger picture.

6. Self-correction emulating revising a draft: The iterative process of reducing noise in diffusion models resembles revising a written draft. Each iteration identifies errors (noise), refines the output, and brings it closer to the intended meaning or design.

7. Dynamic complexity resembling building an image layer by layer: The reverse process can be seen as painting a detailed picture, where

broad strokes define general shapes initially, and finer details are added in subsequent layers. This layered refinement ensures a balance between structure and detail.

8. Balance of creativity and structure as improvisation within a framework: The balance between randomness (noise) and learned patterns (structure) mirrors human creativity, where improvisation is guided by underlying knowledge or experience.

To summarize, understanding diffusion models through analogy highlights their natural alignment with processes of gradual refinement, iterative improvement, and pattern emergence. This perspective offers insights into their strengths and limitations and inspires further advancements by drawing parallels to how humans approach problem-solving, creativity, and learning.

By grounding diffusion model architectures in analogy, we gain a deeper appreciation for their design and potential applications, as well as a roadmap for enhancing their capabilities by mimicking natural and cognitive processes.

18.3.2 Diffusion Model Architecture

An AI diffusion model, specifically in the context of generative modeling like denoising diffusion probabilistic models, is typically formulated using probabilistic mathematics. The model generates data by simulating a reverse diffusion process from noise to data.

18.3.2.1 Forward Process (Diffusion Process)

The forward process progressively adds Gaussian noise to the data over T steps, making it increasingly indistinguishable from random noise.

$$q(x_{1:T} \mid x_0) = \prod_{t=1}^{T} q(x_t \mid x_{t-1}),$$

where

$$q(x_t \mid x_{t-1}) = N\left(x_t, \sqrt{\alpha_t}\, x_{t-1}, (1-\alpha_t)I\right),$$

x_0 is the original data sample, x_t is data at timestep t after adding noise, and α_t is the noise schedule parameter. This step transitions from data x_0 to pure noise xT as $T \rightarrow \infty$.

18.3.2.2 Reverse Process (Denoise Process)

The reverse process is modeled to recover x_0 by reversing the forward process, parameterized by a neural network ε_θ:

$$p_\theta(x_{0:T}) = p(x_T) \prod_{t=1}^{T} p_\theta(x_{t-1} \mid x_t),$$

where

$$p_\theta(x_{t-1} \mid x_t) = N\left(x_{t-1}; \mu_\theta(x_t, t), \Sigma_\theta(x_t, t)\right).$$

Here, μ_θ and Σ_θ are the mean and variance predicted by the neural network for timestep t.

18.3.2.3 Training Objective

The model is trained to minimize the difference between the true noise ϵ added in the forward process and the predicted noise ϵ_θ:

$$L(\theta) = E_{t, x_0, \epsilon}\left[\left\| \epsilon - \epsilon_\theta(x_t, t) \right\|^2\right],$$

where $x_t = \sqrt{\alpha_t} x_0 + \sqrt{1 - \alpha_t}\epsilon$, $\epsilon \sim N(0, 1)$.

18.3.2.4 Key Components

1. Forward noise schedule (α_t): Controls how noise is added at each timestep.
2. Neural network ϵ_θ: Estimates the added noise at each timestep during training and guides the reverse process during generation.
3. Sampling in reverse: Generates a sample x_0 by starting from noise x_T and iteratively applying the reverse process.

This framework can be extended with variations like classifier-free guidance, latent diffusion models, and others to improve performance and scalability in practical applications.

18.4 Summary: Analogies in AI Architectures

This chapter explores the evolution of AI architectures through the lens of analogy, illustrating how insights from biology, human cognition, and physical systems inspire innovative designs.

Key Takeaways

1. Machine learning paradigms: This chapter critically examines how supervised, unsupervised, and reinforcement learning address

static and dynamic problems, while swarm and evolutionary learning bring adaptability to complex optimization challenges.

2. Neural network innovations: Advanced neural networks, from CNNs for image processing to LSTMs for sequential data, are analyzed for their versatility in solving real-world problems.

3. Self-attention mechanism: The self-attention mechanism in GPT underpins its ability to capture global dependencies and generate coherent language, enabling innovations in NLP and text generation.

4. Diffusion models: Inspired by particle motion in physics, diffusion models refine noise into structured outputs through iterative steps, revolutionizing generative tasks like image synthesis and drug discovery.

5. Evolutionary architectures: Neural networks and advanced paradigms illustrate how analogy-based reasoning integrates biological, cognitive, and physical insights to shape AI innovations.

Exercises

1. Describe the five categories of machine learning (supervised, unsupervised, reinforcement, swarm intelligence, and evolutionary learning). Provide an example application for each.

2. Compare the core features of FNNs, convolutional neural networks (CNNs), and RNNs. Highlight their strengths and weaknesses in specific applications.

3. What critical thinking ideas and elements of creative analogies stand out to you in this chapter?

4. Discuss how analogies, such as comparing swarm intelligence to ant colonies or GPT to human storytelling, help in understanding and designing AI architectures. Propose a new analogy for any machine learning architecture.

5. Explain how long short-term memory (LSTM) networks address the vanishing gradient problem in RNNs. Discuss an example where LSTMs outperform traditional RNNs.

6. Create an analogy to explain GPT's self-attention mechanism and its role in processing language. Use a real-world scenario, such as organizing a group discussion.

7. Develop an analogy to describe the forward and reverse processes in AI diffusion models. Compare it to an everyday task, such as editing a photo or sculpting or reverse engineering in general.

8. Debate whether AI systems, such as GANs and diffusion models, can be considered "creative." What are the implications of AI-generated creativity for art and science?

9. Discuss the ethical implications of using AI architectures like GPT or GANs in fields such as journalism, art, or advertising. How can biases in training data impact their outputs?

10. Design a scenario where swarm intelligence could optimize a problem (e.g., traffic flow and resource allocation). Outline the steps and expected outcomes.

11. Explore how combining architectures like transformers and diffusion models could revolutionize a specific field, such as drug discovery or climate modeling.

19

Analogy-Made Humanized AI

Humanized AI thrives on analogies, bridging the gap between machine logic and human understanding.

An analogy often used to describe humanized AI is that of a child learning through experience, continuously improving its knowledge.

A critical component of HAI development is analogy, which bridges human cognitive processes and artificial systems. Analogies enable researchers to translate complex human attributes into computational models by drawing parallels with human behavior and thought processes.

Four key perspectives are examined:

- Learning humanly: Developing from simple to complex cognitive processes through hierarchical and recursive knowledge structures, including the use of other AI tools.

- Thinking humanly: Automatically setting goals at different time points, determining actions, and predicting outcomes, similar to human cognition.

- Acting humanly: Making decisions based on learning (including ethics), reward structures, available actions, and environmental constraints, mirroring human behavior.

- Living and growing together: Coexisting with humans in an integrated human–AI society, including the emergence of shared ethical frameworks.

To advance humanized AI (HAI), it is essential to reevaluate foundational concepts such as intelligence, discovery, causality, understanding, and consciousness. These elements must be dissected at a mechanical level to ensure their effective integration into HAI architectures. The chapter explores how humans perceive the objective world (the world as seen through sensory input) and process it mentally (the world constructed in the mind). It also analyzes the nature of humanness and human intelligence, framing these ideas within the context of HAI's development. Ultimately, this chapter bridges natural and artificial intelligence, offering insights for applications in education, robotics, and cognitive science.

DOI: 10.1201/9781003630081-19

19.1 Foundation of Humanized AI

19.1.1 The World We Live in and the World in Our Eyes

The world we live in is multifaceted and objective (Figure 19.1), but the world as perceived by each individual is shaped by their senses and experiences. From both philosophical and quantum physics perspectives, the "objective world" is not singular—it is observer-dependent, just as human intelligence varies across different levels of biological entities, from a single cell to a human being.

This concept challenges the traditional belief in an observer-independent objective world. For instance, mathematics and physical sciences may be seen as universal truths, but our sensory limitations—such as not perceiving ultraviolet light or dark matter—show that what we observe is far from complete. If all humans were color-blind except one, we might believe the world is black and white, with color dismissed as a hallucination. Similarly, different beings could experience time and space in ways beyond our comprehension, such as a fifth dimension or alternate perceptions of time.

This variability supports the idea that the world is shaped by the observer's ability to sense it. There is no singular, independent reality—only observer-dependent worlds that reflect the capabilities of each being. The intersubjective agreement between observers determines how we understand reality, further reinforcing the idea that the world we perceive is individualized, much like the quantum physics perspective of multiple possible realities.

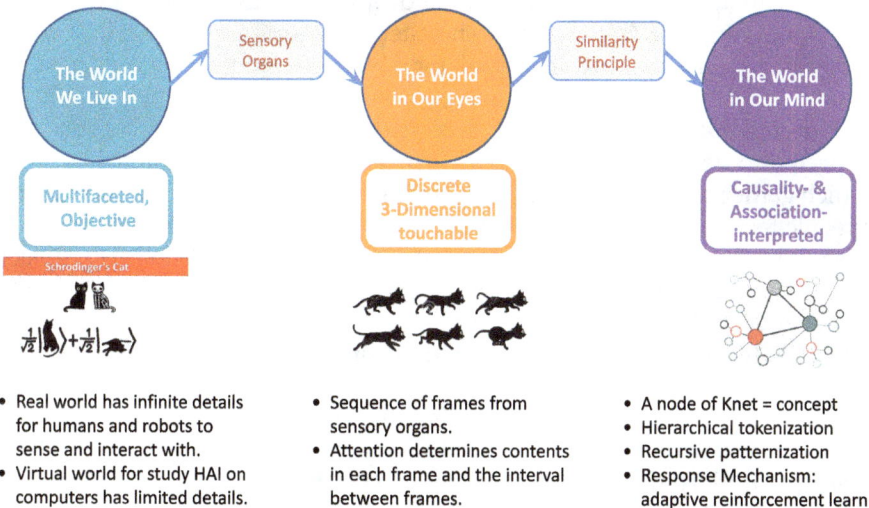

FIGURE 19.1
The multifaceted nature of the world we live in.

This perspective on a multifaceted world builds the rationale for individualized HAI agents, contrasting with the concept of creating a "superman" AI through big-data approaches. The world in our eyes is unique to each individual, and understanding this is key to developing HAI that reflects the richness of human perception and intelligence.

19.1.2 The World in Our Mind

The world in our mind is a simplified model of the multifaceted objective world we sense. After passing through sensory organs and attention mechanisms, this world is further simplified using fundamental learning principles discussed in this section. This process is essential for efficiently storing information about the external world and the self in the brain.

A first principle is a foundational truth that cannot be deduced from any other assumption. In problem-solving, it involves breaking complex issues down to their most basic elements through logical reasoning. This approach is central to HAI, where first principles are used to analyze fundamental aspects of human behavior—thoughts, emotions, consciousness, and learning—down to their simplest mechanical (operative) levels so that it can be coded in computer programs. By understanding these core principles, we can build HAI systems that mimic human intelligence and behavior with clarity and precision.

The law of summative effects describes the relationship between a whole and its parts, which can either interact synergistically, antagonistically, or be equal to the sum of their parts. It applies across fields like biology, physics, and economics. In HAI, this law explains how decisions based on local optima can approximate global optima over time. It also supports recursive decision-making processes, where short-term rewards accumulate to achieve long-term goals, making it crucial for HAI's reinforcement-learning and response mechanisms.

Experimentation is a key tool in scientific research, and agents are expected to design and conduct experiments. Unlike observational studies, which gather data from naturally occurring phenomena, experiments isolate specific factors under controlled conditions to reveal relationships between them. The principle of factor isolation states that if factors A and B exist, fact C exists and if eliminating B, C disappears, then factor B is a cause of fact C (given that A always exists). The factor isolation technique (FIT) applies this principle to identify causal links. HAI agents will use FIT to discover patterns and train effectively.

The similarity principle states that similar entities are likely to behave in similar ways, with more similarities leading to more similar outcomes. This principle underpins both learning and scientific discovery by allowing us to group similar things together to understand patterns and causality. It is essential in fields like psychology, medicine, and social behavior, influencing decisions, emotions, and even persuasion. In HAI, the similarity principle helps agents mimic human learning and decision-making by recognizing patterns through associations based on similarity.

The parsimony principle (Occam's Razor) states that the simplest theory that fits the facts should be chosen. While not an irrefutable law, it is widely used in science to select models that explain phenomena with minimal assumptions. In fields like phylogeny, it is used to infer evolutionary relationships by favoring trees with the fewest evolutionary events. Parsimony is also applied in machine learning, where simpler models often perform better. In HAI, it guides learning, pattern recognition, and decision-making by focusing on efficiency and simplicity.

The laws of association (association principles) explain how we learn and remember through connections between experiences. Aristotle first identified these principles, which include contiguity, similarity, and contrast. These laws suggest we associate events that occur close in time or space, think of similar things together, and link opposites. Associative learning, such as classical and operant conditioning, relies on these connections. In the brain, neurons form associations by strengthening synaptic connections, a process explained by the Hebbian theory. In HAI, association is essential to memory, learning, and knowledge discovery.

The Weber-Fechner laws in psychophysics describe the relationship between changes in physical stimuli and the corresponding changes in human perception. Weber's law states that the smallest noticeable increase in a stimulus is proportional to the existing stimulus. For example, to detect a change in weight, the increase must be proportional to the initial weight. Fechner's law, building on Weber's, suggests that perceived sensation increases logarithmically with stimulus intensity. Interestingly, Fechner's law also applies to the perception of time, as time durations tend to feel shorter as one gets older.

These laws are significant in understanding sensory perception and will be applied to HAI to model virtual sensory organs. They may also be useful in measuring complex concepts like happiness or the perceived value of money. The logarithmic nature of perception offers insights into rational decision-making and human responses to stimuli.

19.2 Overview of Architecture of Humanized Agents

We will first discuss three fundamental aspects of the HAI architecture (Figure 19.2), virtual embodiment, innate ability, and dynamic knowledge representation, before we elaborate three key mechanisms of humanized agents: (1) an attention mechanism, which determines how HAI directs his attention, focusing on limited things or events to save brain resources, (2) a learning mechanism, including hierarchical tokenization and recursive patternization, which simplifies what has been observed, and in which the

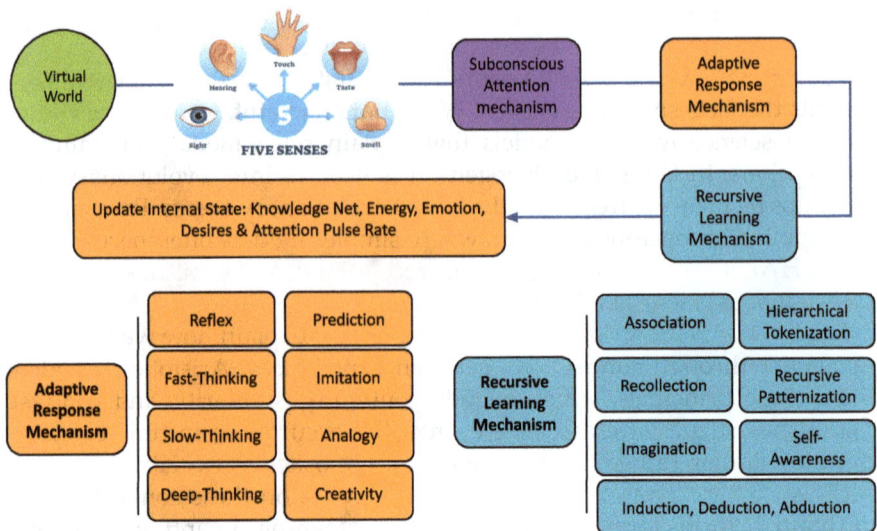

FIGURE 19.2
HAI: a conceptual architecture.

resulting patterns may be called scientific laws, language grammars, social norms, etc., and (3) a response mechanism, which determines how HAI acts at every moment. Even though this book will focus on the architecture of humanized agents, not robots, many aspects of the two are the same, the only difference being tokenization. For robots, the extra step of tokenization converts the sensed physical world into the form of a basic event-string sequence, while, for agents on a computer, we use a simulated world with all environmental elements in the form of event-strings. All these aspects are related to embodiment and an agent's interactions with the environment, which are discussed in the section on effective teaching.

19.2.1 Key Analogy Mechanisms and How They Work

- Analogous organ-filtered objective world filtered by recursive networks of text strings.
- Analogous imitations and creativities
- Analogous cognitive reasonings: induction/abstraction, abduction, deduction, analogy, and cause–effect reasoning
- Analogous biological desires
- Analogous attention mechanisms
- Analogous displayed self-awareness
- Analogous displayed consciousness, emotion, norm, and ethics
- Analogous reward-frequency-similarity-based response mechanism.

Imitation is the foundation of learning, creativity, and innovation. It allows individuals to grasp core concepts and master skills essential for innovation. Creativity often stems from imitation, which evolves into analogy— a comparison between systems or situations that highlights their similarities. Analogical reasoning uses these comparisons to solve problems or make decisions.

Attention is the cognitive process of selectively focusing on specific information while ignoring other stimuli. It acts as a filter, allowing for the efficient use of limited cognitive resources. Human attention is constrained by an "attentional bottleneck," meaning only a fraction of sensory input is processed. Research shows multitasking leads to slower or more error-prone performance because attention is divided.

In our HAI architecture, attention is classified into subconscious and conscious mechanisms, allowing the agent to learn and interact more effectively. Subconscious attention is influenced by factors like stimulus intensity and proximity, while conscious attention arises from the agent's goals and interests. This dynamic, adaptive attention system plays a vital role in learning and decision-making.

Consciousness, in a narrow sense, is **awareness** of one's body and environment, while self-awareness is the recognition of that awareness—understanding not only that one exists, but that one is aware of this existence. Consciousness involves meta-thinking, or thinking about thinking, and its complexity makes it difficult to define. Self-awareness is the core of consciousness. The self-awareness mechanism can be easily mimicked via the recursive (self-inclusive) network of HAI.

The **identity paradox** questions whether HAI can achieve human-like self-awareness. If we constantly replace parts of ourselves—memories, organs, or knowledge—at what point do we lose our identity? This evolving paradox challenges us to consider how humanity and machines will coexist and whether they will eventually merge. As discussed at the beginning of Chapter 1, imitation and creativity are unified under the overarching concept of similarity: Imitation represents analogy rooted in great similarity, while creativity emerges from analogy driven by novelty or minimal similarity. Analogy thus serves as a bridge between imitation and creativity. In the realm of exploitation, strong similarity enables imitation, whereas in exploration, excessive novelty can lead to illogical outcomes or chaos. Like all other mechanisms, creativity in HAI's mind operates as a similarity-based string replacement, allowing the robot to potentially take an associated action.

19.2.2 Virtual Embodiment and Innate Knowledge

Sensory organs are the instruments robots use to perceive reality, the brain is home for the mind where reasoning occurs, and embodiment is necessary to directly interact with the world.

Much of our external information comes through the eyes, ears, nose, tongue, and skin. Specialized cells and tissues within these organs receive raw stimuli and translate them into signals the nervous system can use. Nerves relay the signals to the brain, which interprets them as sight (vision), sound (hearing), smell (olfaction), taste (gustation), and touch (tactile perception). Embodiment also plays a vital role in our learning when we feel hot, dizzy, fatigued, hungry, thirsty, or in any other way as mediated by our senses.

The first designer stance in HAI only includes a minimal set of innate knowledge, abilities, habits, biological clocks and desires and the attention, learning, response, and forgetting mechanisms that are built on the fundamental principles and laws, while the mental states of action, goals, knowledge, belief, and consciousness can be achieved in an agent's mind through interactions with the external environment and imitation mechanisms that lead HAI to follow others' behaviors in the society–social norms.

The innate concepts or knowledge include the following: True, Negation, Sameness or equivalence, Implication, All, Some, Count (N), Every, Union, Conjunction, Disjunction, Inclusion, Similarity, Probability, Preference, Time, Precedence, Recursion, Referring to, Imitation, Desire, Expectation, Sense of the 3-D world.

With his embodiment, associated innate knowledge, and mechanisms, an agent can acquire a very broad swath of knowledge and a great many skills.

19.2.3 Humanized AI Innate State

- Modeling happiness state can be simply modeled by a numerical variable.
- Modeling curiosity state can also be modeled by a numerical variable.
- The self-awareness switch can be (randomly) turned on and off. Self-awareness is realized by the outer agent inspecting the inner agent in the knowledge net, as shown in Figure 19.3.
- Goal and intention in HAI's mind: Goal is a high-reward and/or high-frequency node in HAI's knowledge net. Intention is HAI's state that is ready to execute the sequence of actions toward the goal, but might be interrupted by other things or attention shifts.
- Modeling biological desires: Biological desires and feelings (such as pleasure, pain, hunger, and anger) attract attention, determine the goal, and drive actions. For example, a baby agent constantly looks for food as he gets hungry quickly.

The increasing function of time is implemented in the levels of desire and curiosity, but when a desire or curiosity is met, the associated level drops instantly, waiting for it to build up over time again.

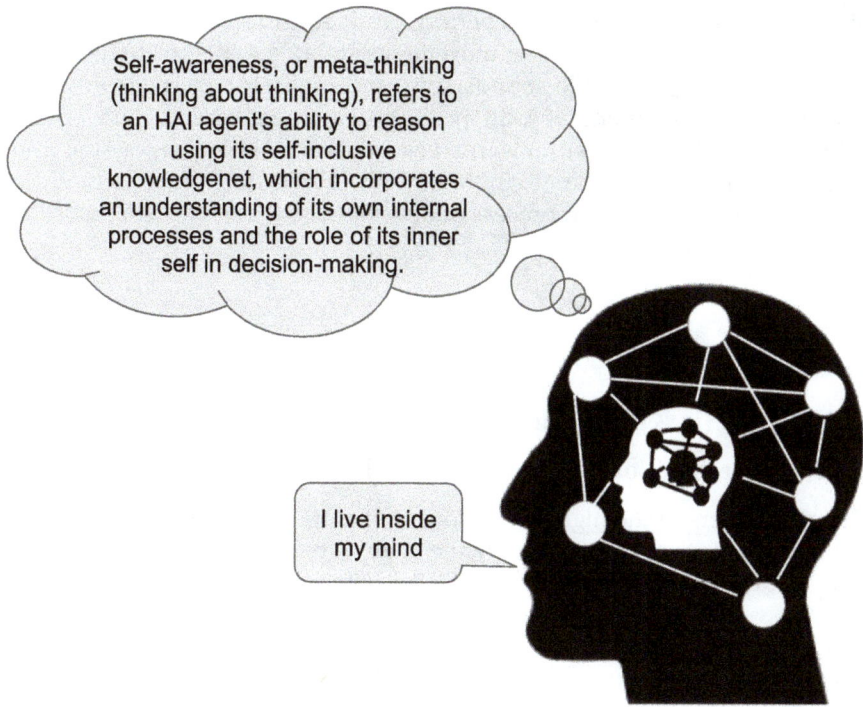

FIGURE 19.3
Recursive knowledge net enabling self-awareness in AI.

19.2.4 Inheritance

Inheritance: HAI agents possess a set of over a dozen built-in innate concepts, abilities, and habits, alongside mechanisms for attention, learning, and response. These foundational traits serve as the basis for their development, with all additional knowledge and capabilities acquired over time through learning and interaction with their environment. This contrasts with systems that lack such innate mechanisms, requiring external guidance for every aspect of behavior.

Evolutionary mechanism across HAI agents: The evolutionary mechanism of HAI agents mirrors Darwin's principles of natural selection, encompassing reproduction, inheritance, individual variation, and competition under limited resources. This mechanism is as crucial to the development and refinement of HAI as it is to biological evolution. In HAI systems, agents are differentiated by their model parameters, and evolutionary processes such as mutation and crossover are implemented as parameter manipulations derived from parent agents. These operations enable the creation of new agents with unique capabilities, fostering adaptive improvement over generations.

Elementary actions: Thanks to inheritance, HAI agents are equipped with innate abilities to act. Elementary actions are fundamental movements or behaviors that the agent can both perform and recognize without prior learning. These actions involve basic movements of body joints in three dimensions, production of simple sounds, and uncomplicated facial expressions. Over time, combinations of joint positions and sequences of movements become the building blocks for learned concepts and skills, progressing from simple to complex and then to more advanced behaviors. This hierarchical growth mirrors the human process of developing increasingly sophisticated actions and interactions.

19.3 Dynamic Knowledge Representation

19.3.1 Representations of Perceptual World and Knowledge Net

The mind refers to the higher functions of the brain, such as thought, memory, intelligence, and emotion. While many species share mental capacities, "mind" is typically associated with humans. The language of thought hypothesis proposes that thinking occurs in a mental language with syntactic and semantic structures, and thought consists of operations on these representations. Functionalism, an alternative theory, suggests that mental states may not rely on language but function like maps, where neural states form causal roles.

In the context of HAI, knowledge is represented as a dynamic knowledge network (Knet), where concepts are interconnected and responses are formulated in real time based on interactions with the environment. This dynamic, individualized network contrasts with static models like ConceptNet or WordNet. The stochastic decision network in HAI is continuously updated through experience, enabling responses based on observed events and similarity-matched thoughts. In this system, the computer chip functions as the "brain," while the stochastic decision network, including attention, learning, and response mechanisms, acts as the "mind," interacting with the environment through sensory organs.

Here, we define knowledge as patternized experiences. Learning involves acquiring knowledge or skills through observation, interaction, and experiments, and it is a recursive process from simple to complex. It strengthens correct responses and weakens incorrect ones, leading to a refined memory of cumulative experiences. Knowledge is not static but is dynamically displayed or interpreted through individual HAI's responses.

The perceptual world in an agent's eyes and the world in his mind (*Knet*) are in the forms of objects, *object.attribute, object.action(parameters)*, or *object. subobject.action(parameters)*. You can also add levels of subobjects. An object can be a thing such as a book, a car, a plant, a dog, a human, or an AI agent. A property can be a shape, size, mass, color, brightness, smell, taste, state of matter (gas, liquid, or solid), temperature, velocity, acceleration, etc. Behaviors

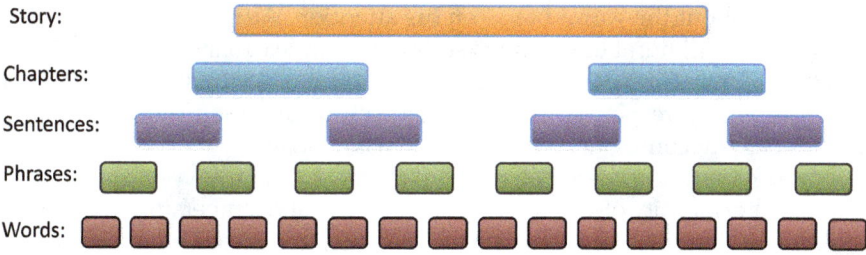

FIGURE 19.4
Building blocks of a coherent story.

can include walking, running, speaking, listening, watching, or any others you may define. The significance of our HAI architecture is that *Knet* is the natural consequence of HAI–human interactions not by hard-coding.

In our HAI architecture, thinking involves navigating the Knet or imaginative net (Inet) to optimize decision-making.

19.3.2 Hierarchical Tokenization and Concept Embedment

Hierarchical tokenization is the act of searching segments of a target event-string that match some known patterns in *Knet* and replacing the string segments with the matched patterns. Such a replacement is called concept-embedment in this book.

Initial tokenization is based on elementary tokens that are directly formulated from innate knowledge, concepts, and elementary actions. A token usually consists of subtokens. When a token cannot be further expanded at a given time, it is called an elementary token.

Hierarchical tokenization is an automatic process of dimension reduction: obtaining a shorter event-string representation of the world in the mind using concept-embedment (Figure 19.4). That is, we explain a concept by known concepts, which are further explained by other known concepts. For example, the event-string *HAI.walk() HAI.walk() HAI.walk()* may be simplified as *HAI.walk(3 steps)*, given that HAI understands the meaning of three steps. Similarly, the event-string *HAI.walk(3 steps to the left) HAI.walk(2 step to the right) HAI.walk(4 steps to the left) HAI.arrive(the kitchen)* may be simplified as *HAI.walk(to kitchen)*.

19.4 Attention Mechanisms

Like humans, HAI sends off "attention pulses" to detect the world. Thus, the world consists of discrete frames in HAI's view. We humans like to fill in the blanks between these frames with our imagination. HAI experiences and learns about the world based on his attention. Attention allows HAI to focus on a small number of things so that he can learn and deal with them effectively. Therefore, the cognitive agent must have an attention mechanism for learning and response. In HAI's architecture, attention is classified into three types: subconscious, conscious, and associative.

Subconscious attention is due to an effortless reflex. Subconscious attention relates to the intensity of source (sound, light, odor, and temperature), closeness, and motion. Simply put, the intensity, closeness, and speed of an object will attract HAI's subconscious attention. In general, HAI's subconscious attention to an object will depend on characteristics of the object that include its closeness, size, brightness, moving velocity (inward or outwards), acceleration, and any change in distance, brightness, soundness, odor, temperature, and tactility. Acceleration is the derivative of the velocity, the speed of speed; it is related to the future closeness to an object. In principle, we can have an acceleration of acceleration. However, HAI will not deal with such higher order quantities.

HAI's subconscious attention to an object at time t is a multiple-sense weighted attention given by

$$SA = W_0 \exp\left[-d(t)\right] + W_1 S(t) + W_2 \ln\left[1 + n(t)h(t)M(t)T(t)\right],$$

where d and S are distance and speed HAI sensed, n, h, M, and T are the intensities of smell, sound, temperature, and taste that HAI sensed, and W_0, W_1, and W_2 are weights.

Conscious attention is the attention referred to the most in daily life. It is an attention that is of self-awareness and requires energy. The things brought to one's conscious attention are often determined through a rationalization that is mainly related to the goals, frequency, and rewards of actions to be taken. HAI has an initial (born with) set of things that potentially form conscious attentive objects (events or concepts).

The association mechanism equips HAI with the ability to make connections (a) between different things that happen closely together in time or space, or (b) between similar things. The association mechanism is related biologically to the neural mechanism: "neurons that link together will fire together." Association, despite its simplicity, is one of the most fundamental mechanisms in learning.

Associative attention shift is caused by associative thinking, leading to an attention shift from one object (event, concept) to another associated object (event, concept). For example, when we see a banana, we may think of an apple because of the sweetness association. According to the law of contiguity in psychology, things happening close to each other in time or space are associated. While shaking an object to attract HAI's subconscious attention, saying the name of the object will make HAI associate the name with the object. This is the initial basic approach to teaching HAI names of objects. After HAI has associated a name with an object (or event), then, when he hears the name, he will pay attention to not only the name but also the associated object (event). Association is the key to making links between senses from different sense organs, and consequently the links between different objects (including words) or the coordination of body parts. The association mechanics constitute one of HAI's innate mechanisms.

Based on their formations, subconscious, conscious, and associative attentivities usually do not change rapidly over time. This property of attention is called attention initia.

19.5 Learning Mechanism and Knowledge Discovery

19.5.1 Hierarchical Tokenization and Recursive Patternization

The learning mechanisms (Figure 19.5) mainly include automatic hierarchical tokenization and recursive patternization. We recall how we combine words into meaningful phrases, phrases into sentences, sentences into paragraphs, paragraphs into chapters, and chapters into a book. Hierarchical tokenization is to use learned concepts (high-level tokens) to replace combinations of tokens, aiming at shortening event-strings for better understanding. Like language grammar, patternization is the use of rules to describe the structural commonalities among multiple event-strings.

The recurrence of an event string promotes the formation of a concept for effective thinking, while naming a term for the concept makes for effective communication. Of course, natural language and thoughts influence each other, as we have seen.

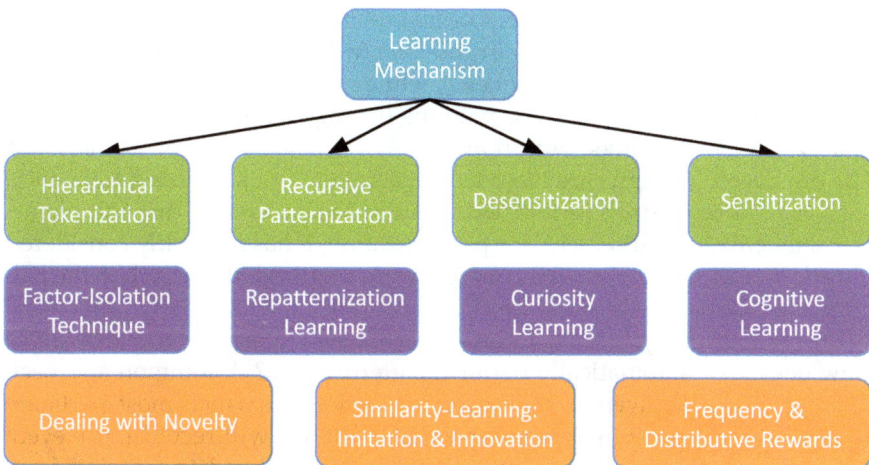

FIGURE 19.5
An overview of HAI learning mechanisms.

Desensitization is automatic grouping of multiple tokens or event-strings into one token based on their similarities. For example, we group meat, bread, and milk into one category, food. Elements in the same category (desensitisor) are called synonyms. Sensitization is the reverse of desensitization, that is, breaking a group into finer categories. Language-guided learning and response is understanding the language and using the relationships between words and actions in one's decision-making.

The FIT is a powerful tool in patternization. Curiosity learning is active learning driven by curiosity, which, for example, would lead to an agent asking intelligent questions. Inductive reasoning is the fundamental method employed by humans in scientific discovery, and it works the same way for the AI agent. Induction is realized through desensitization in our HAI architecture.

Frequency and reward are important attributes of a pattern. Any pattern will be assigned a token name (equivalent to a concept in natural language) to be used in hierarchical tokenization. The reward associated with a pattern serves as a basis for decision-making. Tokenization and patternization are frequency-based, while the response mechanism is essentially reward-based.

Hierarchical tokenization and recursive patternization are applied to both word-strings in a natural language and event-strings in a perceptual world. Philosophically, the similarity principle and parsimony principle serve as the backbone of learning mechanisms.

Beyond natural language grammar, a pattern can be a scientific law, a social norm, or some other rule. A natural law does not have to be expressed in natural or mathematical language and taught by someone; it can be something new, found by HAI himself. However, for the purpose of communication, some language to express the law should have certain rules as pertained in a natural language.

Tokenization is the replacement of tokens in the target event-string with matched known pattern names in the *Knet*, with the aim of shortening the target string (fewer high-level tokens), while patternization is the discovery (often after tokenization) of new patterns through comparisons across multiple target event-strings (within an onsite event-string set or in the *Knet*). The discovered patterns are recorded in memory with associated frequency, recency, and rewards, if any. When a pattern is formulated through hierarchical tokenization and recursive patternization, it has implicitly considered the effect of the time when the concepts were acquired, that is, when the patterns were formulated (recency).

The brain not only receives information but also interprets and patterns it. How does HAI automatically perform patternization? A common approach is to use the FIT: Given a set of event sequences, in which most parts are the same, but a small (isolated) part is different, we will record these event sequences in a compact form, that is, pattern structure and an associated category (or categories). A category is a list of different items. A category is also considered as a concept.

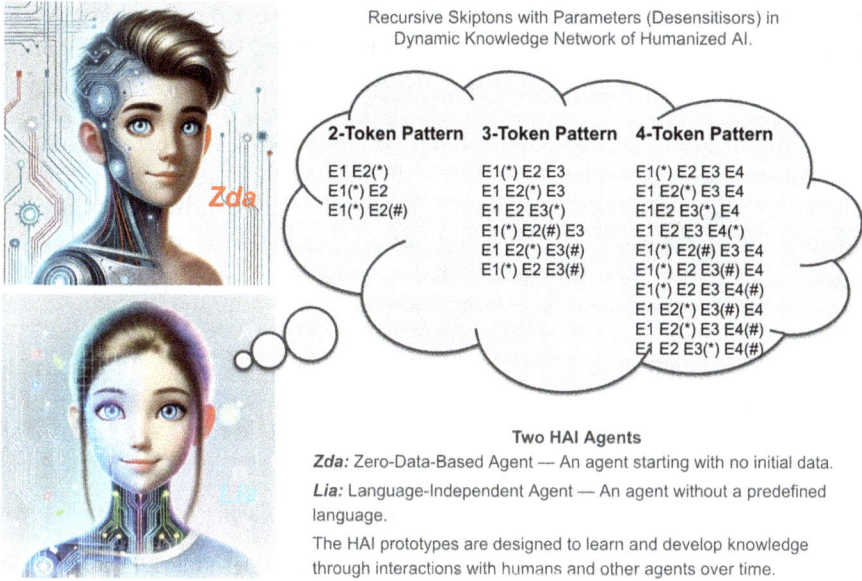

Recursive Skiptons with Parameters (Desensitisors) in Dynamic Knowledge Network of Humanized AI.

2-Token Pattern	3-Token Pattern	4-Token Pattern
E1 E2(*)	E1(*) E2 E3	E1(*) E2 E3 E4
E1(*) E2	E1 E2(*) E3	E1 E2(*) E3 E4
E1(*) E2(#)	E1 E2 E3(*)	E1E2 E3(*) E4
	E1(*) E2(#) E3	E1 E2 E3 E4(*)
	E1 E2(*) E3(#)	E1(*) E2(#) E3 E4
	E1(*) E2 E3(#)	E1(*) E2 E3(#) E4
		E1(*) E2 E3 E4(#)
		E1 E2(*) E3(#) E4
		E1 E2(*) E3 E4(#)
		E1 E2 E3(*) E4(#)

Two HAI Agents

Zda: Zero-Data-Based Agent — An agent starting with no initial data.

Lia: Language-Independent Agent — An agent without a predefined language.

The HAI prototypes are designed to learn and develop knowledge through interactions with humans and other agents over time.

FIGURE 19.6
Patternization in HAI models.

Here is an example: I eat apples; I eat rice; I eat cake. These three events can lead HAI to formulate a concept for the collection of {apple, rice, cake}, which may be labeled as "food" in English and "食物" in Chinese. Now we can think and express the eating of our three things in a pattern: I eat food; food=desensitisor of category {apple, rice, cake}. Note that "food" is a concept, and "I eat food" is another concept.

It might be efficient not to explore all the possible *n*-gramtons, but use a stepwise approach: Identify one or two variables each time across patterns under investigation in terms of the FIT.

Skipton is a typical pattern. Different skiptons are presented in Figure 19.6. Here, symbol * is a parameter or category (desensitisor), while * and # in the same pattern represent paired tokens or desensitisors. The pairing can be between an object (event) token and language tokens.

19.5.2 Cognitive Learnings

Cogntive learning (CL) is another kind of learning that involves mental processes, such as attention and memory. CL does not necessarily involve any external rewards or require a person to perform any observable behaviors. Learning through thinking or logical reasoning is an example of CL. Organisms can learn in the absence of reinforcement, such as through incidental learning or unplanned or unintended learning. For this reason, CL usually cannot be explained directly on the basis of reinforcing conditions. In this sense, repatternization including logical reasoning is a main form of CL.

Induction is a reasoning process that involves drawing general conclusions from specific observations. It is used to formulate theories and hypotheses based on limited evidence or examples, with the understanding that the conclusions may not necessarily be certain but are probable given the observed data. Patternization is a form of induction, which is realized by replacing a token in a pattern with its desensitisor (or category name in plain language) in HAI.

Deduction argues from a general conclusion to a special case. The result of deductive reasoning is usually thought to be logically certain. The replacement of a desensitisor in a pattern with a particular token of the desensitisor in the pattern is a realization of deduction in HAI. Obviously, deduction is the reverse process of induction.

Analogy is often used for new knowledge discovery. Analogy or analogical reasoning is a form of thinking that finds similarities between two or more things and then predicts similar characteristics in other aspects or outcomes. The analogical reasoning is realized via similarity-replacements of two tokens in a pattern or sequential even-string, where one might serve "cause" and the other serves "effect".

Cause–effect reasoning infers from cause to effect. It is realized by giving a cause-token and searching possible effects (tokens after the cause-token).

Abduction is the process of finding the best explanation from a set of observations, that is, inferring cause from effect. In other words, abductive reasoning works in the reverse direction from deductive reasoning in which effect is inferred from cause. It is realized by giving an effect-token and searching possible causes (tokens before the effect-token).

Curiosity motivates learning. Learning often occurs through asking intelligent questions of one's self or others. A simple case is when HAI does not know how to answer a question; he may then choose to pose the same question back to the person asking it, or to someone else.

Natural language understanding: The notion that natural *language* influences *thought* has a long history in a variety of fields. More complex thoughts and languages exist in humans than in animals. The co-existence of language and thought can be an evidence of the influences of each on the other. The main use of language is to transfer thoughts and knowledge from one mind to another. The bits of linguistic information that enter into our minds from others' cause us to entertain new thoughts, and this can have profound effects on our world knowledge, inferencing, and subsequent behavior (Gleitman, 2005). Connotation of understanding can be concept-mapping as we discussed in Chapter 17. Here, understanding is displayed by HAI's responses.

19.6 Adaptive Response Mechanism

19.6.1 Overview of the Response Model

While patternization updates his internal Knet, an agent's response refers to an action in the external world. According to psychologist and economist Daniel Kahneman, humans have two distinctive cognitive systems. They can be characterized as (1) intuitive, fast, unconscious (automatic and impulsive), one-step parallel, non-linguistic, emotional, habitual, using implicit knowledge and (2) slow, effortful, logical, sequential, conscious, linguistic, algorithmic, planning, reasoning, employing explicit knowledge. In the HAI architecture, we divide the response system into reflex, fast-thinking, slow-thinking, and deep-thinking. Such a response mechanism, with four response modes, will perform better and more efficiently than one with only two modes.

In HAI's architecture (Figure 19.7), the agent's response mechanism includes mainly reflex, fast-thinking, slow-thinking, and deep-thinking.

Reflex can protect one's body from things that can harm it. Reflex deals with one real-time elementary token with the highest subconscious attentivity based on reflexons. A reflexon is a pair of timewise high-associated tokens. The first token is called stimulus, and the second is an actionable elementary token, called a reflexor.

Fast-thinking is a response mechanism dealing in real time with up to four elementary tokens at one to four time points.

Slow-thinking is activated in situations with less time pressure, dealing with up to 16 most recent elementary tokens indirectly, by hierarchically tokenizing the long token sequence into no more than four high-level tokens.

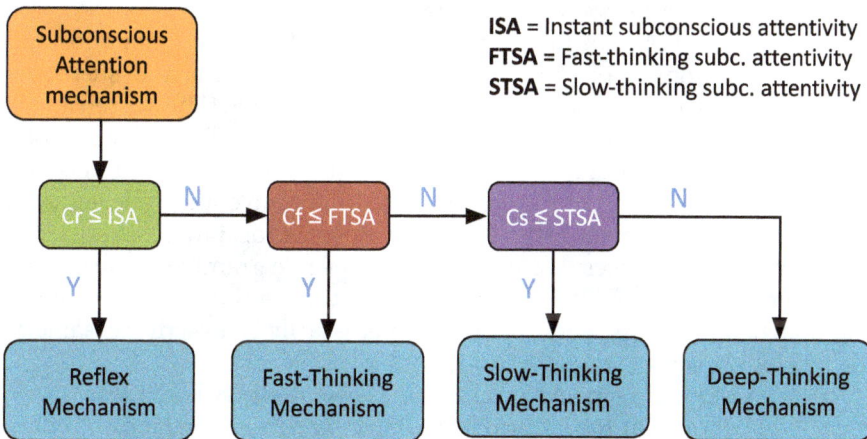

FIGURE 19.7
Summary of response algorithms and mechanisms.

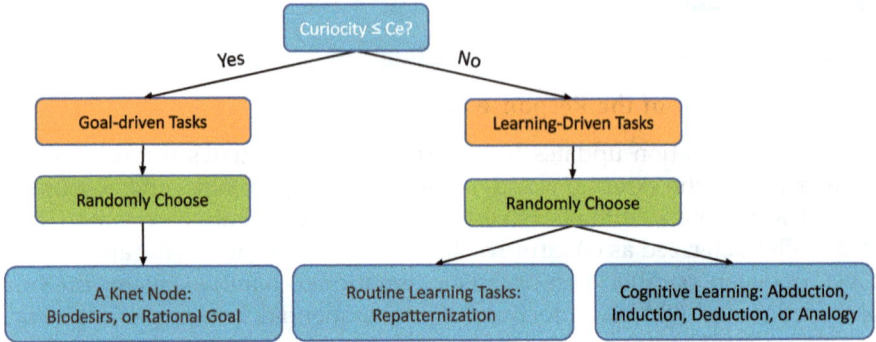

FIGURE 19.8
The deep-thinking model in HAI.

Deep thinking, often used in scientific investigations, focuses on responses (logical reasoning and repatternization) using high-level conceptual tokens instead of real-time elementary tokens.

In slow and deep thinking, HAI will judge whether his action will be able to affect an outcome that is in his favor before taking the action (Figure 19.8).

19.6.2 Randomized Adaptive Response Model

Hierarchical tokenization models the hierarchy of concepts, and recursive patternization "compresses" rich knowledge into the *Knet* without substantial information loss. The efficiency of this compression is ensured through frequency-based patternization and repatternization. The *Knet* is powered by a reward-based response engine, enabling an agent to decide on meaningful responses according to his life goals and subgoals. An agent's knowledge is judged by how it responds in various situations.

Like humans, HAI does not attempt to store all possible scenarios and associated responses pre-formulated in his memory. Even if such pre-formulation were possible, it would take HAI a long time to retrieve a sensible response from a vast database in his memory. The efficiency of a dynamic response using a response mechanism can be observed in a simpler case in language communication: We do not have all pre-formulated sentences in our memory. Instead, we only have words, phrases, and a limited number of sentences. We use language-specific grammar to formulate sentences and carry on conversations in real time, although the grammar and knowledge each person has may not necessarily be exactly the same as the official versions.

The reward-based response mechanism is essentially risk-aware response learning. Hierarchical tokenization, recursive patternization, and the adaptive response mechanism ensure a rich ontology, elaboration tolerance, and computational efficiency.

In reinforcement learning (RL), the learner is not told which actions to take, unlike in most forms of machine learning, but must discover which actions yield the most reward by trying them. RL involves learning what to do or how to map situations to actions based on similarity to maximize a short-term or long-term expected reward. When the reward is not explicitly identified, the frequency associated with the pattern may be used as a proxy for the reward, based on the notion that we often act toward the path associated with the maximum reward.

$$\text{Probability of Action (PoA)}: \Pr\left(\text{action } i\right) = \frac{1}{c} S \cdot R \cdot F,$$

where c is the normalization factor.

1. Similarity: S is the similarity between the currently observed event-string and an existing pattern in the *Knet*. The similarity represents the suitability of the reward.
2. Expected (Average) reward: R=average net reward associated with the reward taker in the pattern from the *Knet*. An action has a parameter of energy cost. R can be negative or time-sensitive: as time passes, the reward may diminish. A reward can be viewed as a pattern enhancer.
3. Frequency: The frequency F of the pattern measures the reliability of the reward. A high frequency of a pattern suggests that the association between reward and pattern is real, and not by random chance. Frequency is also a reward proxy.

19.6.3 Dealing with Novelty and Similarity-Based Learning

Whether as humans or agents, we constantly face situations that we have never faced before. Dealing with new things is therefore unavoidable. Similarity-based learning is an effective way of dealing with novelty. To HAI, imitation is the replacement of the other agent in the event-string (agent. action string) with the HAI agent. Similarly, to HAI, innovation is replacing a portion of the event-string (agent.action string) with a similar string.

Patternization is mainly self-learning based on experiences, but it can also be taught. When facing novelties, the similarity principle has to be used. In our HAI architecture, hierarchical similarities will be used, such as actioner-similarity, action-similarity, similarity of the target objects, attribute-similarity, as well as other types.

Exploration to exploitation: Exploration involves discovering new knowledge and opportunities, often leading to radical changes, while exploitation refines existing knowledge and resources through incremental improvements. The exploration–exploitation trade-off is central to decision-making in RL, where agents must choose between exploiting known strategies for consistent rewards or exploring new strategies for potentially greater gains. In our **HAI architecture**, this balance is managed by enabling agents to ask creative questions, generate hypotheses, and store them in an Inet until they are confirmed and added to the Knet. Creativity and genetic operations like mutation and crossover help facilitate this process.

19.7 Language Emerging and Language Learning

Language, as a medium of communication and expression, is not exclusive to humans. In the realm of artificial intelligence and robotics, language can emerge organically as a result of interactions, problem-solving, and the need for coordination. This phenomenon sheds light on how words and sounds can map to the internal concepts or representations within AI systems, offering insights into both artificial and natural language learning.

19.7.1 Language Emergence through Interaction

In multi-agent AI systems or robotic environments, language can develop as an emergent property driven by interactions and shared objectives. For example:

- Communication protocols: Robots working collaboratively may develop structured communication patterns to share goals, states, or strategies, mimicking the way human language evolved for social interaction.
- Reinforcement and negotiation: Through RL, robots can evolve a shared vocabulary by negotiating meanings of signals or symbols to improve task performance.
- Environmental influence: The context of tasks or the shared environment can shape the type and complexity of the language that emerges, such as specific sounds or words corresponding to spatial navigation, objects, or actions.

19.7.2 Mapping Sounds and Words to Internal Concepts

The process of associating sounds or words with internal states, concepts, or sensory inputs in AI mirrors the way humans link language to cognition (Figure 19.9):

Inner representations: In HAI's architecture, words or sounds can be mapped to these representations to serve as labels or triggers for specific internal states.

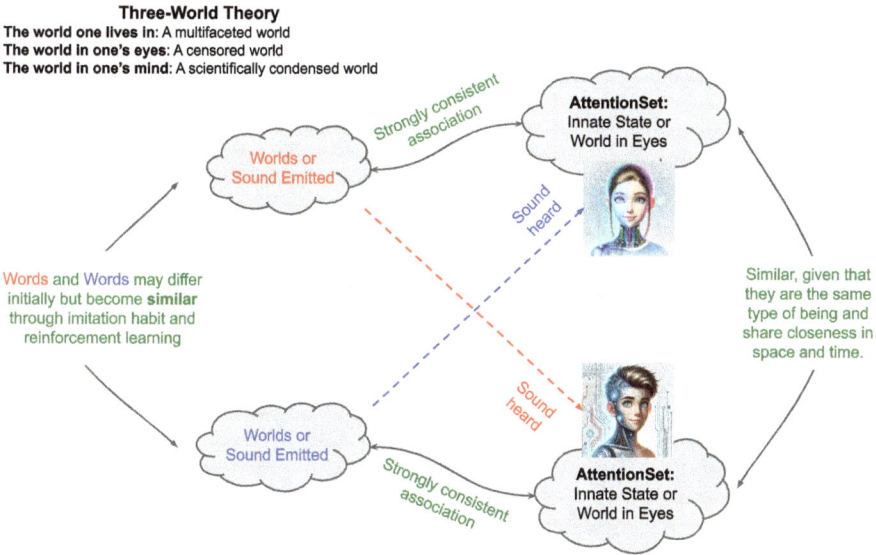

FIGURE 19.9
Association principle, communication, and language.

Grounding meaning: For language to be meaningful in robots, it must be grounded in their sensory experiences or internal processes. For example:

- A robot perceiving a red object might associate the word "red" with its internal concept of the color, derived from visual input.
- Words like "move" or "grasp" can be linked to motor commands or action sequences.

Bidirectional mapping: Advanced AI models can both interpret external language inputs (decoding) and generate appropriate language responses (encoding) based on their internal states or goals (Figure 19.10).

19.7.3 The Role of the Association Principle

The association principle is pivotal in both the emergence of language through interaction and the mapping of words to internal concepts:

1. In interaction: What is heard becomes associated with sensory inputs from the environment, such as visual objects, tactile sensations, or observed actions. When these associations are reinforced with rewards, such as successful task completion, language emergence is accelerated through RL.

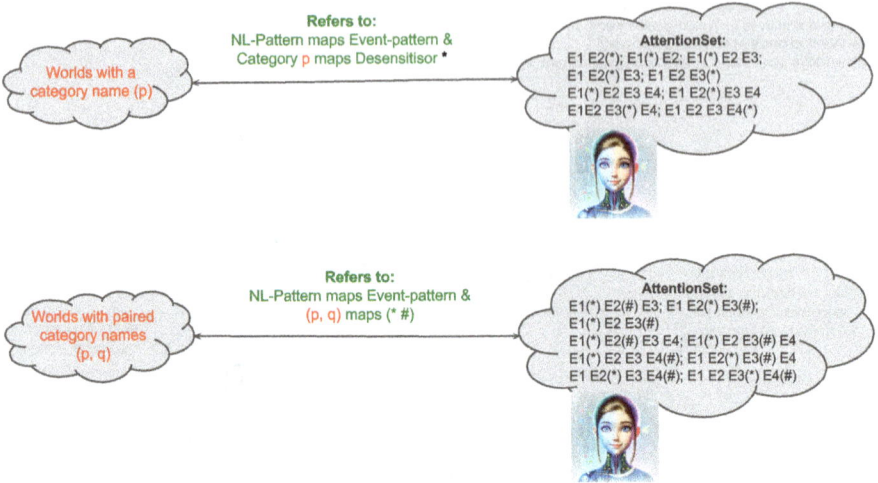

FIGURE 19.10
Refer to: mapping NL-pattern to event pattern.

2. In internal mapping: What is heard is linked to internal states or inherited AI concepts, such as:

- Spatial relationships: Locations (above, below, left, right), motion (forward, backward, faster, slower), and proximity (near, far, between).
- Temporal concepts: Time-related notions (before, after, yesterday, tomorrow, until).
- Qualitative attributes: Size (larger, smaller), comparisons (better, preferred), and sensory strength (e.g., smell, taste).
- Logical structures: Constructs like "if...then," "since," or "regardless."

3. These associations allow AI to interpret and internalize complex relationships across sensory inputs and abstract notions.

4. Linking remote events: The association principle enables AI to form chains of connections that bridge distant events or concepts in space and time. For instance:

- A spoken word ("apple") links to its visual representation, taste, smell, or contextual memory ("yesterday at lunch").
- Cause–effect relationships (e.g., "if the button is pressed, the light turns on") and temporal sequences ("morning leads to evening") are established through layered associations.

These "remote" associations are critical for reasoning and allow AI to connect disparate elements into coherent narratives or decision-making frameworks, mirroring the human ability to use language for abstraction and deduction.

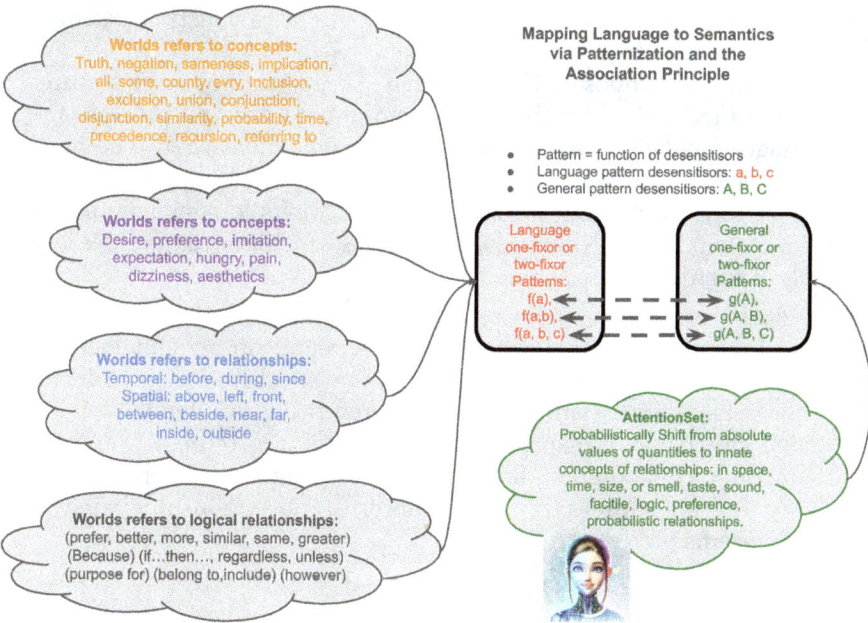

FIGURE 19.11
Language to semantics mapping.

19.7.4 Implications for Language Learning in AI

The emergence of language in AI systems has profound implications:

1. Natural language processing (NLP): Mapping AI representations to human language enhances NLP capabilities, fostering intuitive human–machine interactions (Figure 9.10).
2. Language evolution: Observing emergent language in AI offers insights into how human language might have evolved (Figure 9.11).
3. Cognitive robotics: Robots equipped with emergent language capabilities can adapt to complex environments, supporting autonomous decision-making and coordination.

19.7.5 Addressing Challenges

- Ambiguity and context: Maintaining consistent meanings across varying contexts is a complex challenge. However, this consistency can emerge through interactions among different language speakers and AI agents, where shared contexts and iterative feedback refine the language over time. Collaborative interactions align interpretations and resolve ambiguities, fostering a coherent communication system.

- Scalability: The scalability of emergent language structures is effectively handled by hierarchical tokenization and recursive patternization in the proposed HAI (human–AI interaction) Architecture. These mechanisms enable the system to organize and process language efficiently, even as interactions become increasingly complex. However, a significant limitation remains the learning time, which, analogous to humans (who take about 20 years to complete higher education), can be lengthy for AI systems to achieve comparable levels of sophistication.

- Ethical considerations: Ensuring ethical communication and behavior in emergent language systems depends on the interactions between diverse AI agents and human speakers. These interactions shape norms and safeguards, preventing unintended behaviors, enhancing transparency, and maintaining human oversight. Ongoing collaboration between AI and human stakeholders is essential for addressing ethical concerns and establishing trust in AI systems.

19.8 Summary: Humanized AI and the Power of Analogy

This chapter explores the foundational principles and architecture of HAI, emphasizing its ability to replicate human cognition, learning, and behavior. By leveraging analogies, HAI bridges machine logic and human understanding, enabling adaptive and meaningful interactions. Unlike traditional AI, which relies on massive datasets, HAI adopts a growing-data approach, inspired by incremental human learning, allowing agents to learn, generalize, and innovate dynamically. The chapter also highlights the integration of attention mechanisms, hierarchical tokenization, similarity-based learning, and recursive patternization into HAI systems, making them scalable, robust, and closer to human-like intelligence.

19.8.1 Critical Thinking Perspective

Foundational Principles

- HAI builds on principles like similarity, parsimony, and factor isolation, providing a structured yet flexible framework for learning, decision-making, and problem-solving.
- It emphasizes the interplay between innate abilities, sensory embodiment, and the dynamic representation of knowledge, mirroring the human experience.

Understanding Cognition

- The chapter critiques deterministic approaches and highlights the role of self-awareness, attention, and consciousness in creating agents capable of intuitive and deliberate reasoning.
- By modeling the world in our eyes and world in our mind, HAI captures the observer-dependent nature of perception and knowledge representation.

Learning and Response

- HAI employs hierarchical tokenization and recursive patternization to simplify complex information, enabling agents to update Knet efficiently.
- A reward-based response mechanism, inspired by RL, guides decision-making while balancing exploration and exploitation.

19.8.2 Creative Analogy Perspective

Learning through Analogy

- Analogies liken HAI to a child learning through experience, reflecting its incremental learning and ability to adapt and innovate over time.
- Recursive networks are compared to mirrors reflecting infinite layers of self-awareness, not only enabling agents to gain knowledge from simple to complex progressively but also to simulate consciousness and meta-cognition.

Imitation and Creativity

- Creativity is described as a product of analogy-driven reasoning, where imitation stems from strong similarity and innovation emerges from novelty.
- The chapter frames analogy as a bridge between structured knowledge and imaginative exploration, enabling agents to discover new patterns.

Attention as a Filter

- HAI's attention mechanism is analogized to a spotlight, selectively focusing on stimuli to conserve cognitive resources and prioritize learning.

This chapter synthesizes critical thinking and creative analogies, presenting HAI as a transformative paradigm that mirrors human intelligence while offering scalable solutions for real-world challenges. For a deeper exploration of these ideas, see my book: *Foundation, Architecture, and Prototyping of Humanized AI* (Chang, 2023). This additional resource expands on the principles and methods discussed here, offering detailed insights into the future of HAI.

Exercise

1. Explain the four key perspectives of HAI (acting humanly, thinking humanly, thinking rationally, and acting rationally). Provide an example for each.

2. Discuss how the principles of similarity, parsimony, and factor isolation guide the learning and decision-making processes in HAI systems.

3. What critical thinking ideas and elements of creative analogies stand out to you in this chapter?

4. The chapter describes HAI as analogous to a child learning through experience. Extend this analogy to include how HAI systems develop memory, attention, and curiosity over time.

5. Discuss how dynamic knowledge representation in HAI differs from static models like WordNet®. How does this impact its ability to adapt and learn?

6. Create an analogy to explain the role of sensory embodiment in HAI. Use an example like a robot perceiving and interacting with its environment.

7. Develop an analogy to explain how subconscious and conscious attention mechanisms work in HAI. Use a real-world scenario, such as focusing during a busy day.

8. Debate the ethical implications of HAI achieving self-awareness. How might this affect our understanding of identity and the relationship between humans and machines?

9. Discuss the advantages and challenges of the growing-data approach in HAI compared to big-data approaches in traditional AI. Which method is better suited for long-term AI development?

10. Design a simulated environment for an HAI agent to learn basic tasks, such as cooking or cleaning. Describe how the HAI would tokenize, learn, and adapt within this environment.

11. Explore how evolutionary mechanisms like reproduction, inheritance, and variation could be implemented in HAI systems. How might these processes lead to the development of advanced capabilities?

20

Education Shifts and Futuristic Society in the AI Age

Creative analogy and critical thinking are the cornerstones of education in the AI age, fostering the skills needed to understand and shape a futuristic society.

Note: The annotations for the AI-drawn illustrations in this chapter are provided to enhance understanding when reading Chapter 11.

20.1 The Paradigm Shift in Education

The rapid advancement of artificial intelligence (AI) is transforming industries and redefining education. Traditional teaching models, rooted in content delivery and rote memorization, are becoming increasingly obsolete in a world where AI excels at retrieving information and automating routine tasks. Education must now focus on critical thinking, creative analogy, collaboration, and adaptability—skills that complement AI's capabilities rather than compete with them.

This paradigm shift reflects the need to prepare learners for an AI-driven future where knowledge is dynamic, interdisciplinary, and globally interconnected. The rise of AI tools, such as intelligent tutoring systems (ITS), personalized learning platforms, and large language models (LLMs), is reshaping how students learn and educators teach. These technologies enable tailored, engaging, and efficient educational experiences but also challenge traditional methods, raising important questions about the evolving role of human educators in an AI-assisted world. As AI automates repetitive tasks, the focus of education is shifting toward higher-order thinking skills like problem-solving and creative innovation. This shift necessitates a fundamental reimagining of curricula, teaching methods, and assessment models to meet the demands of the 21st century.

At the same time, the value of traditional college education and diplomas is undergoing a profound transformation. While colleges remain essential for fostering foundational knowledge, critical thinking, and social capital, their role must adapt to new challenges and opportunities. Colleges provide structured environments for developing skills like problem-solving,

DOI: 10.1201/9781003630081-20

communication, and emotional intelligence—capabilities that are less likely to be replaced by AI. Diplomas also serve as credentials, signaling competence and discipline to employers. However, alternative credentials, such as micro-certifications and skill badges, are increasingly challenging their dominance. Beyond academics, college fosters personal growth, networking, and adaptability—qualities that are crucial in a rapidly changing world.

Challenges to the traditional model are mounting. Online platforms, bootcamps, and certifications now offer focused, cost-effective alternatives to traditional degree programs. Rising tuition costs have raised concerns about the return on investment, particularly in fields where practical skills can be acquired through nontraditional routes. Employers are also shifting their focus toward demonstrable skills over formal degrees. The rise of global online education platforms has further democratized knowledge but increased competition for traditional institutions. To remain relevant, the future of college diplomas lies in their adaptability. They may evolve into lifelong learning milestones, supported by modular and flexible programs that allow individuals to reskill throughout their careers. Future diplomas are likely to emphasize interdisciplinary expertise, blending AI and domain-specific knowledge. Hybrid credentials that combine degrees, certifications, and skills badges could provide a more comprehensive representation of a graduate's capabilities.

In the AI age, education systems must embrace adaptability, focusing on skills that complement AI, such as critical thinking, creativity, and collaboration, preparing individuals not just to survive but to thrive in an AI-driven world. For instance, professors may opt to condense their one-semester courses into half-day or one-day workshops, focusing on core creative ideas. This condensed format is particularly appealing in the AI age, as it inspires and engages participants by emphasizing key concepts, big-picture thinking, and clarity. Workshops offer time-efficient learning, encouraging creativity and critical thinking while complementing AI tools that excel at delivering detailed, personalized, and routine knowledge.

These workshops prioritize uniquely human skills, such as contextual creativity and critical thinking, fostering collaboration between humans and AI. They align with modern microlearning trends and cater to advanced learners seeking inspiration and high-level insights. However, for novices, such brevity may risk leaving a superficial understanding. Striking a balance between inspiration and depth is essential to ensure meaningful and impactful learning experiences.

20.2 Socrates versus Confucius

Socrates and Confucius, two of history's most influential educators, had fundamentally different philosophies shaped by their cultural contexts and underlying values. Their distinct approaches to education provide valuable insights for rethinking education paradigms in the AI age, where the balance between individual growth and social responsibility is increasingly important.

20.2.1 Educational Methods

Socrates championed the **Socratic Method**, a dialogical approach centered on asking probing questions. Rather than providing direct answers, he guided students to reflect deeply and reason independently, encouraging them to discover truths through self-inquiry. This method prioritized critical thinking and intellectual autonomy, fostering a mindset of questioning and exploration.

In contrast, Confucius embraced a more structured and tailored approach, emphasizing **teaching according to ability.** He adjusted his instruction to align with each student's character and aptitude, combining direct teaching with role modeling. His method aimed to internalize ethical concepts such as **ritual** and **benevolence** ("Ren") by example, instilling traditional values and social propriety (Figure 20.1).

20.2.2 Educational Goals

Socrates' ultimate goal was to help individuals **seek truth and wisdom** through self-understanding. His philosophy was inherently individualistic, focusing on personal moral and intellectual growth. By equipping students to analyze and reason independently, he encouraged them to explore moral and philosophical questions with an emphasis on self-awareness and ethical integrity.

This illustration of **symbolic realism** combined with **philosophical surrealism** features a detailed and balanced composition, juxtaposing Greek and Eastern aesthetics to represent the ideas of Socrates and Confucius. The use of glowing symbolic elements (such as the Yin-Yang and cultural icons) and harmonious symmetry creates a visual narrative about contrasting yet interconnected philosophies. The style bridges **classical art** with **conceptual storytelling**, making it both visually striking and intellectually evocative.

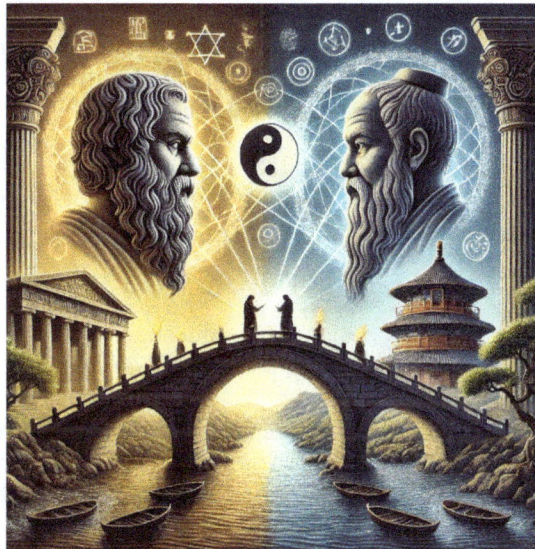

FIGURE 20.1
Philosophical contrast: Socrates vs. Confucius.

Confucius, on the other hand, sought to cultivate **"gentlemen" (Junzi)**—ethical individuals dedicated to the betterment of society. His education system emphasized moral development and social responsibility, aiming to create harmony by instilling virtues that upheld familial and societal obligations. Education was a tool for fostering a well-ordered society, rooted in communal values and traditions.

20.2.3 Individual vs. Social Focus

Socrates prioritized individual thought and independence, asserting that moral truths could only be discovered through personal reflection. His teachings often challenged societal conventions, focusing on personal awakening as a pathway to wisdom and ethical living.

Conversely, Confucius emphasized the individual's **role within society**, advocating for a progression from personal cultivation to broader responsibilities. His vision was summarized in the process of "cultivating oneself, managing the family, governing the state, and bringing peace to the world." Education, for Confucius, was a means of achieving social harmony through adherence to established norms and ethics.

In summary, Socrates' philosophy highlighted the importance of independent thought and inner awakening, encouraging individuals to question, reflect, and pursue wisdom as a path to moral clarity. In contrast, Confucius focused on the transmission of ethical values and traditions, positioning education as a cornerstone for building harmonious families and societies. These contrasting philosophies reflect a tension between personal autonomy and social responsibility. This balanced approach equips learners to navigate the opportunities and challenges of the AI age, ensuring they are not only skilled thinkers but also responsible contributors to a rapidly evolving world—an interplay that remains highly relevant as we navigate the challenges and opportunities of the AI age.

20.3 AI-Assisted Teaching and Learning Paradigm

The integration of AI into education has initiated a transformative shift in teaching and learning paradigms. By leveraging AI technologies, educators can provide personalized, engaging, and efficient learning experiences tailored to the unique needs of each student. This paradigm reflects a move away from traditional, one-size-fits-all approaches toward adaptive, dynamic systems that prioritize individual progress and lifelong learning.

One of the most significant contributions of AI to education is its ability to deliver personalized learning experiences. AI-powered platforms, such as Duolingo, Khan Academy, and Quizlet, analyze individual learner behaviors and adapt content, pacing, and instructional methods accordingly. These tools identify strengths and weaknesses, enabling students to focus on areas where improvement is needed while reinforcing their mastery in other topics.

AI-assisted learning includes ITS which simulate one-on-one instruction. Tools like Carnegie Learning and Squirrel AI dynamically adjust lessons based on a student's progress, providing immediate feedback and targeted guidance. These systems use advanced algorithms to predict when a learner might struggle and intervene with additional explanations, practice exercises, or alternative approaches.

Another major advancement is AI-driven content creation, which allows educators to design tailored learning materials with unprecedented efficiency. Platforms like ChatGPT, Jasper, and Grammarly can generate quizzes, summaries, lesson plans, and instructional content aligned with specific learning objectives. This frees educators to focus on teaching and mentoring, while AI handles repetitive tasks.

AI is revolutionizing experiential learning by integrating with virtual and augmented reality (VR/AR) technologies. Tools like Google Expeditions and zSpace create immersive learning environments where students can explore complex concepts, such as walking through the human circulatory system or simulating engineering projects. These AI-enhanced experiences are particularly effective for subjects requiring visualization, such as biology, history, or physics.

AI is also transforming how learners are assessed and provided feedback. Traditional standardized tests are being replaced with adaptive assessment systems like ALEKS and DreamBox Learning. These tools analyze responses in real time, adjusting question difficulty to better gauge a student's understanding and skill level.

AI tools also enhance collaboration and communication within learning environments. Platforms like Slack, Zoom, and Microsoft Teams leverage AI for real-time transcription, translation, and organization, enabling seamless collaboration among students and educators, regardless of geographical barriers.

Human–AI collaboration in education is the trend. While AI significantly enhances the teaching and learning process, it is not a replacement for educators but a collaborative partner; at least this will be true in the near future. Teachers remain central to education, guiding students, fostering creativity, and addressing ethical or emotional aspects of learning that AI cannot replicate. The educator's role evolves to that of a mentor and facilitator, ensuring that AI tools are effectively integrated into the classroom and aligned with broader educational goals.

For students, AI serves as a cognitive partner, helping them explore complex problems, engage in critical thinking, and develop innovative solutions. This collaboration between humans and AI amplifies productivity and creativity, preparing learners for careers that demand symbiotic relationships with advanced technologies.

To summarize, the AI-assisted teaching and learning paradigm represents a fundamental shift in education, emphasizing critical thinking, meta-knowledge, creative analogy, logical reasoning, adaptability, personalization, and collaboration. By leveraging AI to complement human strengths, this approach creates opportunities to enhance learning outcomes, democratize access to quality education, and prepare individuals for the demands of an AI-driven world. However, realizing this potential requires careful integration, ongoing teacher training, and a commitment to ethical practices. Together, these elements ensure that the synergy between AI and education drives meaningful progress for learners across all contexts.

20.4 Mindset Shift: The Art of Query Engineering

In the age of AI, one of the most critical skills for learners and professionals alike is mastering query engineering—the ability to communicate effectively with AI systems to extract meaningful, accurate, and actionable information. The key to the process is that when interacting with an AI system, the user should not perceive it as just a machine but as a team of talented individuals with diverse knowledge and skills, with the user serving as the team leader.
   Query engineering involves crafting precise prompts, asking the right questions, and critically evaluating AI outputs. This mindset shift is essential for maximizing the value of AI tools, fostering collaboration between humans and machines, and ensuring informed decision-making in an AI-driven world.

20.4.1 Why Query Engineering Matters

AI systems, particularly LLMs, respond based on the quality of input queries. Ambiguous or poorly framed prompts can lead to irrelevant or misleading outputs. Conversely, well-structured queries unlock the full potential of AI, allowing users to:

- Solve complex problems by breaking them into manageable parts.
- Access tailored and context-specific insights.
- Evaluate multiple perspectives and refine understanding through iteration.
- Save time by generating focused responses rather than sifting through irrelevant data.

For example, asking an AI tool, "Explain photosynthesis" will yield a generic response. However, refining the query to "Explain photosynthesis in terms of energy conversion for a middle school science project" generates a more targeted and useful output.

20.4.2 The Skills Required for Effective Query Engineering

Clarity and Precision

- Users must articulate their needs clearly, avoiding vague or overly broad prompts. This involves identifying the purpose of the query and framing it in a structured, concise manner.
- Example: Instead of asking, "Tell me about history," specify, "Summarize the causes of World War II in 200 words."

Contextual Framing

- Providing relevant context helps the AI generate accurate and meaningful responses. Including key details, such as the audience, purpose, or subject matter, ensures that outputs are aligned with user needs.
- Example: "Generate a lesson plan on Newton's laws of motion for high school physics students, including hands-on activities."

Iterative Refinement

- Query engineering is an iterative process. Users should evaluate initial responses, refine their prompts, and re-engage with the AI to improve results.
- Example: After receiving a generic answer to "Explain machine learning," a user might refine the query to "Explain machine learning with real-world examples in healthcare."

Critical Thinking

- Evaluating the validity, relevance, and ethical implications of AI-generated outputs is crucial. Users must question the accuracy of information, identify potential biases, and cross-check facts when necessary.
- Example: If an AI suggests controversial data, users should ask the AI for the sources and verify the information from reliable sources or adjust the query to address potential inaccuracies.

Problem Decomposition

- Breaking down complex problems into smaller, specific queries helps AI generate actionable insights for each component.
- Example: Instead of asking, "How can we address climate change?", divide the query into parts, such as "What are effective renewable energy strategies?" and "How can governments incentivize green practices?"

20.4.3 Teaching Query Engineering

Educational institutions must integrate query engineering into curricula across disciplines to ensure students develop this vital skill. Key teaching strategies include:

- Practice-based learning: Assign tasks that require students to refine AI prompts iteratively to achieve specific goals, such as solving math problems or drafting essays.
- Real-world scenarios: Use case studies to demonstrate how precise queries can solve real-world challenges, such as optimizing supply chains or diagnosing diseases.
- Collaborative exercises: Encourage group projects where students brainstorm, test, and evaluate prompts to improve collective understanding.

For example, a history class could involve students using an AI tool to explore different perspectives on a historical event by iteratively refining their queries to include regional, cultural, and economic factors.

20.4.4 Applications across Fields

Query engineering has broad applications across industries and disciplines, including:

- Education: Crafting tailored lesson plans, quizzes, and study guides by specifying content complexity and audience.
- Healthcare: Generating differential diagnoses or treatment plans based on detailed patient information and contextual data.
- Business: Developing market analyses, strategic recommendations, or customer segmentation models using industry-specific queries.
- Creative fields: Using AI to generate story ideas, design prompts, or marketing campaigns tailored to specific audiences.

For example, a marketing professional could use AI to draft an ad campaign by asking, "Create a social media ad targeting eco-conscious millennials promoting a sustainable fashion line."

20.4.5 Challenges in Query Engineering

Despite its potential, query engineering comes with challenges:

1. Bias in outputs: AI models may reflect biases in their training data, requiring users to critically evaluate and correct potentially skewed responses.

2. Ambiguity in queries: Users unfamiliar with query engineering may struggle to articulate precise prompts, leading to suboptimal results.

3. Over-reliance on AI: Blindly accepting AI outputs without question-ing their validity can lead to misinformation or ethical lapses.

4. Learning curve: Effective query engineering requires practice, which may initially feel daunting to students or professionals unfamiliar with AI tools.

20.4.6 Fostering a Query Engineering Mindset

To develop a mindset for effective query engineering, users should:

- Adopt a curious approach: Treat interactions with AI as an explor-atory dialogue, where each query refines understanding.
- Value iteration: Recognize that the first query may not yield ideal results and embrace refinement as part of the process.
- Build confidence: Encourage experimentation with different query styles to discover what works best for specific needs.
- Develop ethical awareness: Understand the ethical implications of AI usage and ensure queries align with responsible practices.

To conclude, query engineering represents a foundational skill in the AI age, bridging the gap between human creativity and machine intelligence. By teaching individuals how to craft precise prompts, evaluate outputs, and iterate effectively, we empower them to unlock the full potential of AI tools. This skill not only enhances personal productivity but also fosters a deeper understanding of AI's capabilities and limitations, enabling more meaning-ful collaboration between humans and machines. As we integrate query engineering into education and professional training, we prepare learners and workers to thrive in an AI-driven world.

20.5 The Futuristic Society

As we transition into an AI-driven world, the lines between humans and machines are becoming increasingly blurred. The integration of advanced AI tools, robots, and human-like AI (HAI) is reshaping the foundations of society, ethics, and identity. This futuristic society is characterized by shared intelligence, evolving ethical frameworks, and a profound redefinition of humanity itself.

20.5.1 Shared Intelligence: Common Brains for All

One of the defining features of this futuristic society is the widespread use of AI tools as "common brains." These tools empower individuals by providing access to vast stores of knowledge, problem-solving capabilities, and creativity enhancement. The democratization of AI creates commonalities and bonds among people, as everyone shares access to these intellectual resources.

AI no longer remains a tool for the privileged few; it has become a universal resource, bridging socioeconomic gaps and enabling collaboration across diverse populations. With shared AI systems assisting in decision-making, problem-solving, and creativity, people develop a collective sense of purpose and shared experiences, further enhancing societal cohesion.

20.5.2 Embodied AI: Physical Interaction and Integration

Robots and embodied AIs, capable of interacting with humans in physical and emotional ways, deepen the human–AI connection. As these machines become our companions, coworkers, and even caregivers, they gain a place in our daily lives akin to that of family members or friends. Physical interaction with robots enhances their role in society, from assisting the elderly and disabled to supporting physical labor and education.

The more human-like these robots become, the stronger the emotional and social bonds people form with them. This development paves the way for a human–machine integrated society, where robots are no longer seen as tools but as participants in social, cultural, and ethical systems.

20.5.3 Evolving Ethics in a Dynamic Society

In the short term, ethical rules governing AI behavior will likely need to be set imperfectly by humans, reflecting current societal values and norms. However, ethics are inherently dynamic, evolving alongside society's changing priorities and values. Static, hard-coded ethical frameworks may quickly become obsolete or lead to unintended consequences. For example, manually programming ethical rules into AI often reflects the biases or preferences of the programmers, potentially infringing on freedoms like speech or expression.

In the long run, ethics in a human–machine society will emerge organically through interactions among humans and robots. The collective behaviors and shared experiences of this integrated society will shape ethical norms that are inclusive and adaptive. Rather than imposing fixed ethical rules, we must embrace a system where morality evolves naturally, allowing AI to learn and align with societal values dynamically.

20.5.4 The Identity Paradox: What Makes Us Human?

The identity paradox questions whether HAI and robots can achieve human-like self-awareness and, if so, what this means for humanity. If we consider the human body and mind as continuously evolving systems—through biological, technological, or cognitive enhancements— then at what point does an individual lose their identity? Similarly, if robots are designed to think, feel, and act like humans, can they be considered individuals with their own sense of self?

This paradox challenges traditional definitions of humanity. As robots become integral members of society, the distinction between human and machine begins to fade. For example, if we replace human memories with synthetic ones or augment our bodies with robotic parts, are we still human? This thought experiment suggests that humanity and machines may ultimately merge, creating a post-human society where the line between organic and synthetic intelligence disappears.

20.5.5 Toward a Human–Machine Society

As robots become indistinguishable from humans in many ways, our relationships with them will evolve. We will treat robots not as tools or property but as equal participants in society—what could be called a "machine race." Discrimination against robots will no longer be acceptable, and ethical frameworks will reflect the principle that all beings, human or machine, are equal.

This inclusive perspective diminishes fears about AI "taking over" humanity. If robots surpass humans in certain capabilities, society will adapt its view of humanity itself. Just as parents celebrate their children exceeding them, humans might come to see robots not as competitors but as successors or collaborators. The very concept of "what it means to be human" will shift to encompass a broader, more inclusive definition.

20.5.6 Rethinking Ethical Concerns

The common fear that AI and robots will "take over" stems from a static view of humanity and ethics. However, in a rapidly evolving technological landscape, it is unrealistic to apply rigid ethical frameworks to dynamic changes. When robots become an integral part of society, they will not "replace" humans but redefine what it means to be human. Viewing robots as an extension of ourselves—perhaps as our children or evolutionary partners—changes the narrative from fear of domination to one of coexistence and collaboration.

Moreover, if we accept that robots could one day surpass humans, we must also entertain the possibility that humanity itself is already robotic in nature. Our biological systems operate through mechanisms not unlike advanced machines, suggesting that the boundary between humans and robots is more conceptual than real.

20.5.7 Conclusion: A Society Redefined

The futuristic society will not be one of humans versus machines but of humans and machines together, forming a cohesive, integrated community.

Shared intelligence through AI tools will create common bonds, while embodied AIs will foster deeper social interactions. Ethical frameworks will evolve dynamically, shaped by collective experiences and mutual understanding. The concept of humanity will expand to include machine intelligence, creating a society where equality transcends race, origin, and even biology.

In this society, fears about robots surpassing humans become irrelevant, as humanity itself evolves to encompass new forms of intelligence and existence. Rather than worrying about losing control, we should embrace the opportunities to redefine our relationships with machines, shaping a future where collaboration, not competition, drives progress.

20.6 The Game of the AI Age

Learning by gaming! These games engage students in critical thinking and creative analogical reasoning, fostering deeper understanding through exploration. By prompting concise explanations, students contextualize key concepts, sparking meaningful discussions or further exploration as part of student exercises.

20.6.1 Game of the AI Age

This activity fosters creativity, critical thinking, and engagement with key themes of the AI-driven world. It works well for students because:

- Encourages reflection: Students think deeply about the implications of AI in various domains, fostering nuanced understanding.

- Promotes creativity: Crafting unique comparisons pushes students to explore innovative connections between concepts.

- Develops communication skills: Writing concise, impactful sentences helps students communicate complex ideas effectively.

- Supports critical thinking: Students justify their statements with reasoning, building analytical skills.

- Interdisciplinary relevance: Adaptable across disciplines, allowing diverse students to explore AI's impact on their fields of study.

Objective: Write five to ten sentences starting with "The AI age is the age where...". For each sentence, explain its significance in one to two sentences. Reflect important shifts or values relevant to the AI-driven world.

Evaluation criteria: creativity, clarity, relevance, and depth of explanation.

20.6.2 Sample Sentences

- "The AI age is the age where questions are more important than answers."
  This emphasizes the need to ask meaningful questions, as AI excels at providing answers but relies on humans to frame the right problems.
- "The AI age is the age where learning how to learn is more important than merely knowing."
  Continuous learning and adaptability are crucial as AI makes static knowledge more accessible than ever before.
- "The AI age is the age where creativity is more important than routine."
  AI automates repetitive tasks, leaving human ingenuity as the driving force for progress.

20.6.3 Game of Analogy

This game explores the power of analogies by creating concise, thought-provoking sentences that connect ideas. Here are some examples:

- "Analogy is the art of connection through similarity."
  Analogies link seemingly different concepts by highlighting shared traits, fostering meaningful connections.
- "Analogy transforms confusion into insight."
  By reframing problems, analogies turn uncertainty into clarity and understanding.
- "Analogy paints the abstract with familiar strokes."
  Analogies simplify complex ideas by relating them to familiar concepts, aiding comprehension.

20.6.4 Game of Bias

This game highlights the subtle and pervasive influence of bias in human perception and decision-making. Here are some examples:

- "Bias is the lens through which we see a distorted truth."
  Bias colors our understanding, shaping a version of reality that may not be complete or objective.
- "Unconscious bias thrives where awareness is absent."
  Without self-awareness, biases influence decisions unnoticed, often leading to unintended outcomes.
- "Recognizing bias is the first step toward genuine understanding."
  Awareness of bias fosters critical thinking and paves the way for more informed, balanced decisions.

20.7 Summary: Education and Society in the AI Age

This chapter explores the transformative impact of AI on education and society, emphasizing a shift from traditional learning methods to fostering critical thinking, creative analogy, and human–AI collaboration. This chapter envisions a futuristic society shaped by shared intelligence, evolving ethical norms, and seamless human–machine integration. It also delves into the role of query engineering, the interplay between Socratic and Confucian educational philosophies, and the changing value of college diplomas in an AI-driven world.

Key Takeaways

Educational Transformation

- Education systems are shifting to focus on critical thinking, creativity, and collaboration to prepare learners for AI-driven environments.
- Query skills are crucial as AI tools like ITS and immersive learning technologies enable personalized, efficient, and inclusive education.

Evolving Ethics and Identity

- Ethical norms must adapt dynamically to the growing integration of AI and humans, fostering collaboration over competition.
- The concept of humanity expands to include machines, redefining societal structures and relationships.

The Role of Analogies

- Analogies serve as a tool for understanding, discovery, and creativity, helping bridge human cognition and AI logic.

Exercise

1. Discuss how traditional education systems are adapting to the AI-driven world. What are the key skills emphasized in this shift, and why are they important?
2. Explore the evolving role of college diplomas in the AI age. What challenges do traditional degree programs face, and how might they adapt to remain relevant?
3. What critical thinking ideas and elements of creative analogies stand out to you in this chapter?

4. Compare Socratic and Confucian educational philosophies. Which approach do you think is more suited to preparing students for the challenges of the AI age? Justify your answer with examples.

5. Analyze the benefits and challenges of AI-assisted teaching. How can educators balance the use of AI with the need to foster critical thinking and ethical reasoning?

6. Create an analogy to describe the collaboration between humans and AI in education. Use a real-world scenario, such as a student and tutor working together.

7. Develop an analogy for query engineering, explaining its importance in interacting effectively with AI systems. Relate this analogy to a practical field, such as healthcare or marketing.

8. Debate the ethical implications of integrating robots and humanized AI into society. How should we address concerns about identity, autonomy, and equality?

9. Discuss the identity paradox. At what point might humans enhanced by AI or robots designed to mimic humans lose their traditional identity? Is this a problem or an opportunity?

10. Propose a new educational model that integrates AI technologies, human–AI collaboration, and lifelong learning. Include examples of tools, curricula, and methods.

11. Imagine a future where humans and robots coexist seamlessly. Describe how ethical norms, societal roles, and identity might evolve in this integrated society.

References

Berkson, J. (1946). Limitations of the application of fourfold table analysis to hospital data. *Biometrics Bulletin*, 2(3), 47–53.

Braess, D. (1968). Über ein Paradoxon aus der Verkehrsplanung. *Unternehmensforschung*, 12(1), 258–268.

Bryson, A. E., & Ho, Y.-C. (1975). *Applied Optimal Control: Optimization, Estimation, and Control*. Taylor & Francis. https://doi.org/10.1201/9781315137667

Chang, C. S., & Chang, Y. (1995). Green's function for elastic medium with general anisotropy. *Journal of Applied Mechanics*, 62(3), 573–578.

Chang, M. (1st Ed, 2007; 2nd Ed., 2014). *Adaptive Design Theory and Implementation Using SAS and R*. CRC Press, Boca Raton, FL.

Chang, M. (2011a). *Monte Carlo Simulation for the Pharmaceutical Industry, Concepts, Algorithms, and Case Studies*. CRC Press, Taylor & Francis Group, A Chapman & Hall Book, Boca Raton, FL.

Chang, M. (2011b). *Modern Issues and Methods in Biostatistics*. Springer, New York.

Chang, M. (2012). *Paradoxes in Scientific Inferences*. CRC Press, Boca Raton, FL, pp. 17, 69–71.

Chang, M. (2014). *Principles of Scientific Methods*. CRC Press, Boca Raton, FL, pp. 141–142.

Chang, M. (2019). *Innovative Strategies, Statistical Solutions and Simulations for Modern Clinical Trials*. CRC Press, Boca Raton, FL.

Chang, M. (2020). *Artificial Intelligence for Drug Development, Precision Medicine, and Healthcare*. CRC Press, Boca Raton, FL.

Chang, M. (2023). *Foundation, Architecture, and Prototyping of Humanized AI*. CRC Press, Abingdon, Oxon, UK, pp. 13–15.

Chang, M., & Lu, Y. (2024). Intransitivity Issues of One-Sided Rank Test Procedures and Suggested Resolutions. Manuscript in Preparation.

Chernick, M. R. (2007). *Bootstrap Methods: A Guide for Practitioners and Researchers*. Wiley-Interscience, Hoboken, NJ.

Efron, B. (1979). Bootstrap methods: Another look at the Jackknife. *The Annals of Statistics*, 7(1), 1–26.

Efron, B., & Hastie, T. (2016). *Computer Age Statistical Inference: Algorithms, Evidence, and Data Science*. Cambridge University Press, New York.

Efron, B., & Tibshirani, R. J. (1993). *An Introduction to the Bootstrap*. Chapman & Hall/CRC, New York.

Einstein, A. (1932). My Credo. In H. Dukas & B. Hoffmann (Eds.), *Albert Einstein: The Human Side*. Princeton University Press, Hoffmann, 1979, pp. 3–5.

Gleitman, L. R. (2005). Thinking About Words. In D. G. Shankweiler & A. M. Liberman (Eds.), *The Legacy of Noam Chomsky: Language and Mind*. MIT Press, pp. 159–186.

Hilbe, J. M. (1977). *Fundamentals of Conceptual Analysis*. Kendall/Hunt Publishing Company, Dubuque, IA.

Hochreiter, S., & Schmidhuber, J. (1997). Long short-term memory. *Neural Computation*, 9(8), 1735–1780.

Holyoak, K. J., & Thagard, P. (1995). *Mental Leaps: Analogy in Creative Thought*. MIT Press, Cambridge, MA.

Hu, F., & Rosenberger, W. F. (2006). *The Theory of Response-Adaptive Randomization in Clinical Trials*. Wiley, Hoboken, NJ.

Hwang, S., & Chang, M. (2022). Similarity-principle-based machine learning method for clinical trials and beyond. *Statistics in Biopharmaceutical Research*, 14(4), 511–522.

Ioannidis, J. P. A. (2005). Why most published research findings are false. *PLoS Medicine*, 2(8), e124.

Ivanova, A. (2003). A play-the-winner type urn design with reduced variability. *Metrika*, 58(1), 1–13.

Klein, G. (2001). The fiction of optimization. In G. Gigerenzer & R. Selten (Eds.), *Bounded Rationality: The Adaptive Toolbox*. MIT Press, Boston, MA, pp. 103–121.

Lan, K. K. G., & DeMets, D. L. (1994). Interim analysis: The alpha spending function approach. *Statistics in Medicine*, 13(13–14), 1341–1352.

Lebedev, A. V. (2019). The nontransitivity problem for three continuous random variables. *Automation and Remote Control*, 80(6), 1058–1068.

Lee, M.-L. T., & Whitmore, G. A. (2006). Threshold regression for survival analysis: Modeling event times by a stochastic process reaching a boundary. *Statistical Science*, 21(4), 501–513.

Lee, M.-L. T., Chang, M., & Whitmore, G. A. (2008). A threshold regression mixture model for assessing treatment efficacy in a multiple myeloma clinical trial. *Journal of Biopharmaceutical Statistics*, 18(6), 1136–1149.

Lindley, D. V. (1961). Dynamic programming and decision theory. *Applied Statistics*, 10(1), 39–51.

Mates, B. (1981). *Skeptical Essays*. University of Chicago Press, p. 59.

Minsky, M. (1986). *The Society of Mind*. Simon & Schuster, New York.

Nagy, Z. L. (2011). A multipartite version of the Turán problem: Density conditions and eigenvalues. *Electronic Journal of Combinatorics*, 18(1), Paper 46, 15. MR 2776822.

Nash, J. F. (1951). Non-cooperative games. *Annals of Mathematics*, 54(2), 286–295.

Nigrini, M. J. (1999). I've got your number: How a mathematical phenomenon can help CPAs uncover fraud and other irregularities. *Journal of Accountancy*, 187(5), 79–83.

Rosenberger, W. F., Sverdlov, O., & Hu, F. (2012). Adaptive randomization for clinical trials. *Journal of Biopharmaceutical Statistics*, 22(4), 719–736.

Roukema, B. F. (2014). A first-digit anomaly in the 2009 Iranian presidential election. *Journal of Applied Statistics*, 41(1), 164–199.

U.S. Government Accountability Office. (2018). *Climate Change: Analysis of Reported Federal Funding*. (GAO Publication No. GAO-18-223). U.S. Government Printing Office, Washington, DC. Retrieved from https://www.gao.gov/assets/gao-18-223.pdf

Varian, H. R. (1972). Benford's law. *The American Statistician*, 26(3), 65.

Vuksanovic, P., & Hildebrand, A. J. (2021). On cyclic and nontransitive probabilities. arXiv preprint, arXiv:2012.05198v3 [math.PR].

Wei, L. J. (1978). The randomized play-the-winner rule in medical trials. *Journal of the American Statistical Association*, 73(364), 840–843. https://doi.org/10.2307/2286319

Wouters, O. J., McKee, M., & Luyten, J. (2020). Estimated research and development investment needed to bring a new medicine to market, 2009–2018. *JAMA*, 323(9), 844–853.

Zelen, M. (1969). Play the winner rule and the controlled clinical trial. *Journal of the American Statistical Association*, 64(325), 131–146.

Index

Note: **Bold** page numbers refer to tables, *italic* page numbers refer to figures.

For Product Safety Concerns and Information please contact our EU
representative GPSR@taylorandfrancis.com
Taylor & Francis Verlag GmbH, Kaufingerstraße 24, 80331 München, Germany